THE INSTRUCTOR'S GUIDE

for

HUMAN FACTORS IN FLIGHT

First published 1995 by Ashgate Publishing

Published 2016 by Routledge
2 Park Square, Milton Park, Abingdon, Oxon OX14 4RN
711 Third Avenue, New York, NY 10017, USA

Routledge is an imprint of the Taylor & Francis Group, an informa business

This *Instructor's Guide* accompanies the *Student Workbook* and both relate largely to the main textbook:
Human Factors in Flight, Second Edition, 1993; (c) Frank Hawkins 1987, (c) Uniepers 1993; published by Avebury Aviation/Ashgate Publishing Company.

For details of the *Student Workbook* please contact the publisher.

The Author acknowledges the support and comments of colleagues at various campuses of the Embry-Riddle Aeronautical University, in particular Professors Don Hunt, James Cooper and Shannon Trebbe of the Daytona Beach, FL campus and Dr. Ronald Clark of the Extended Campus extension at Sky Harbor Airport in Phoenix, AZ.

The development of this material is an ongoing process, and the Author and Publisher welcome comments and suggestions. These will be properly acknowledged and credits given when appropriate in further printings.

ISBN 13: 978-0-291-39832-1 (pbk)

INTRODUCTION

This *Instructor's Guide* is designed to help you present the concepts in the *Human Factors in Flight* textbook. The unit lecture note format has been used to organize the material and provides performance objectives, discussion questions with answers, and references to the pages in the text where the source material is found. At the end of each unit's lecture notes are pages containing large print summary material for the creation of overhead transparencies to support the concepts presented in the lecture notes. The discussion questions in the *Instructor's Guide* are numbered to coincide with the unit questions in the *Student Workbook*.

A Master Question File containing objective questions on each unit is also provided in the Appendix to enhance the instructor aids.

The *Instructor's Guide* is also available direct from the Author in Word Perfect on 3 1/2 inch disk.

Table of Contents

Appendix:

Master Question File

HUMAN FACTORS IN AVIATION SAFETY
(SF 320)
LECTURE NOTES

DAY 02   Background to Human Errors

PO: 1.   *Describe the general impact of human factors on the aircraft accident record in recent years.*

1.   Why is the study of Human Factors so important in the promotion of aerospace safety?   (HF p 10)

-        A vast majority (75%) of aircraft accidents are a result of inadequate performance on the part of the human.

2.   How does the human record look in the **worldwide commercial jet** fleet accident record?

   (From Boeing's 1993 Statistical Summary)

	1959 - 1993	LAST 10 YEARS	
Flightcrew as a Primary Factor All Phases	64.6%	58.2%	(p 14)
Flightcrew as a Primary Factor Final Approach & Landing	78.9%		(p 15)

3.   What impact does the human have on the accident record of the **U.S. air carriers**?

   (From 1991 NTSB Annual Review p 18)

Broad Cause/Factor	All accidents		Fatal Accidents	
	1991	1986-1990	1991	1986-1990
Pilot	40.7%	34.7%	25.0%	39.3%
Other Person (not aboard)	37.0%	39.6%	50.0%	57.1%
Other Person (aboard)	3.7%	22.2%	.0%	7.1%

4.   What extent is the human involved in the accident record of **U.S. general aviation** aircraft?

(From 1992 NTSB Annual Review)

Broad Cause/Factor	All accidents (p 22) 1992	1987-1991	Fatal Accidents (p 34) 1992	1987-1991
Pilot	78.3%	82.3%	83.7%	89.3%
Other Person (not aboard)	8.7%	9.0%	10.1%	10.2%
Other Person (aboard)	0.3%	0.6%	0.4%	0.2%

5.   How does "Personnel" fit in as a broad cause or a factor in the 1992 **general aviation** aircraft accidents?

(From 1992 NTSB Annual Review p 39)

- 87.7%   = Personnel

- 83.7%   = Pilot

6.   What do all these statistics boil down to in terms of how many accidents have the human as the cause?

U.S. Air Carrier Pilot as a Broad Cause Factor 1986 - 1990	39.3%
Worldwide Commercial Jet Fleet All Phases of Flight Flightcrew as a Factor 1983 - 1993	58.2%
Worldwide Commercial Jet Fleet Final Approach & Landing Phase Flightcrew as primary factor 1959 - 1993	78.9%
General Aviation "Personnel" as a Broad Cause Factor in 1992	87.7%

PO: 2.   *Give a brief historical perspective on the development of human factors.*

7.   When and with who did Human Factors as a technology start?

(HF p 16)

- 1880s -1890s with Taylor and the Gilbreths time and motion studies in industry

8.   What were the important milestones in the first century of human factors development?

(HF pp 16-18)

    a.   Hawthorne Works of Western Electric from 1924 - 1930

       -       Work effectiveness could be favorably influenced by psychological factors not directly related to the work itself.

       -       "The Hawthorne Effect"

    b.   The Psychology Laboratory at Cambridge constructed a cockpit research simulator during 1940 - 1945.

       -       The "Cambridge Cockpit" concluded that skilled behavior was dependent to a considerable extent on the design, layout and interpretation of displays and controls.

    c.   Aviation psychology centers were initiated at Ohio State and Illinois Universities during 1940s.

    d.   Establishment of ergonomics or human factors as a technology by the founding of the Ergonomics Research Society in the UK in 1949.

    e.   Recognition that basic education in human factors was needed throughout the industry.

       -       A two week course "Human Factors in Transport Aircraft Operations" was started at Loughborough University, England in 1971.

       -       A short course established at USC in the US.

       -       By 1978 KLM provided the first "Human Factors Awareness Course" for large scale, low-cost, in-house indoctrination of staff in basic Human Factors principles.

f.	The 20th Technical Conference of International Air Transport Association (IATA) in late 1975.	(HF pp 19-20)

-	The turning point in official recognition of the	importance of Human Factors in air transportation.

-	Two significant messages from the IATA Technical Conference:

(HF p 19)

1-	Something was amiss related to the role and performance of man in civil aviation.

2-	A basic human factors gap existed in air transport.

g.	NASA's establishment in 1976 of the confidential Aviation Safety Reporting System (ASRS).

h.	The 1977 Tenerief ground collision of two 747s killing 583 people which resulted entirely from a series of Human Factors Deficiencies.

i.	In 1982 the UK set up the Confidential Human Factors Incident Reporting Programme (CHIRP).

9.	When did serious interest get started in generating a greater awareness of human factors amongst those responsible for design, certification and operation of aircraft?
(HF p 11)

- In the 1970s

10.	What did the International Air Transport Association (IATA) conclude about human factors in their 20th Technical conference in 1975 at Istanbul?
(HF p 11)

-	"the wider nature of human factors and its application to aviation seem still to be relatively little appreciated. This neglect may cause inefficiency in operation or discomfort to the persons concerned; at worst it may bring about a major disaster."

11.	What are the two broad principles that must be accepted to achieve better application of human factors in civil aviation?
(HF p 12)

a.	Questions about the role of man within complex systems are technical questions requiring professional expertise.

b.   Adequate resources must be allocated to the design and management of man-hardware-software system.

PO: 3.   *Define the meaning of Human Factors*

12.   What is Human Factors about?

(HF p 20)

-      **People** in their working and living environment.

-      **People** in their relationships with:

-      machines and equipment
-      procedures
-      the environment around them
-      other people

13.   What might be some key words in a good definition of human factors?

(Taken from *Human Factors for General Aviation*, p 1-2)

-      people interact
-      environments
-      pilot performance
-      design of cockpits
-      functions of the organs of the body
-      effect of emotions
-      interaction and communication

14.   What is the applied technology of human factors concerned with?

(HF p 20)

-      How to optimize the relationship between people and their activities by the systematic application of the human sciences, integrated within the framework of systems engineering.

15.   What are the twin objectives of human factors applications?

(HF p 20)

a.   Effectiveness of the system. (Safety & Efficiency)

b.   The well-being of the individual.

16. What does the term "ergonomics" mean?

(HF p 20)

- The study of man in his working environment.

17. How is human factors different from ergonomics?

(HF p 20)

- It has a wider meaning

- It encompasses some aspects of human performance and systems interfaces which would generally not be considered in the mainstream of ergonomics.

18. What were some of the early misconceptions about human factors?

(HF p 21)

- That it was a branch of medicine

- Possibly because of the earliest problems being physiological

19. With what problems, in addition to the physiological, must human factors be concerned?

(HF p 21)

- Human behavior and performance   (Psychological)

    - Decision making and the cognitive processes

    - Design of controls and displays

    - Flight deck and cabin layout

    - Communication

    - Software aspects of computers, maps, charts, and documentation

    - Refinement of staff selection, training, and checking

(next page)

20.   How can you describe human factors as a "multi-disciplinary" technology?

(HF pp 20-21)

-       It takes in many sources of information from the human sciences:

    a.   Psychology

    b.   Physiology

    c.   Anthropometry and biomechanics

    d.   Biology and chronobiology

    e.   Genetics

    f.   Statistics

THE HUMAN FACTOR
IN
THE ACCIDENT RECORD

U.S. Air Carrier
Pilot as a Broad Cause Factor   39.3%
1986 - 1990

Worldwide Commercial Jet Fleet
All Phases of Flight   58.2%
Flightcrew as a Factor
1983 - 1993

Worldwide Commercial Jet Fleet
Final Approach & Landing Phase   78.9%
Flightcrew as primary factor
1959 - 1993

General Aviation "Personnel"
as a Broad Cause Factor   87.7%
in 1992

FIRST CENTURY MILESTONES
IN HUMAN FACTORS DEVELOPMENT

1924-30   -   "The Hawthorne Effect"

1940-45   -   The "Cambridge Cockpit"

1940s   -   Aviation psychology centers at Ohio
                State and Illinois Universities

1949   -   Founding of the Ergonomics Research
                Society in the UK

1971   -   "Human Factors in Transport Aircraft
                Operations" was started at
                Loughborough University, England

1975   -   20th Technical Conference of
                International Air Transport Association
                (IATA)

1976   -   NASA's establishment of the
                confidential Aviation Safety Reporting
                System (ASRS)

1977   -   Tenerief ground collision of two 747s

1978   -   KLM provided the first "Human
                Factors Awareness Course"

1982   -   UK set up the Confidential Human
                Factors Incident Reporting Programme
                (CHIRP)

INTERNATIONAL AIR TRANSPORT ASSOCIATION
HUMAN FACTORS CONCLUSIONS

"the wider nature of Human Factors and its application to aviation seem still to be relatively little appreciated. This neglect may cause inefficiency in operation or discomfort to the persons concerned; at worst it may bring about a major disaster."

TWO BROAD PRINCIPLES
OF HUMAN FACTORS

1. Questions about the role of man within complex systems are technical questions requiring **professional expertise.**

2. **Adequate resources** must be allocated to the design and management of man-hardware-software system.

WHAT IS HUMAN FACTORS ABOUT?

- People in their working and living environment.

- People in their relationships with:

    - machines and equipment

    - procedures

    - the environment around them

    - other people

HUMAN FACTORS:

(Human Factors in General Aviation, p 1-2)

- people interact

- environments

- pilot performance

- design of cockpits

- functions of the organs of the body

- effect of emotions

- interaction and communication

THE APPLIED TECHNOLOGY OF HUMAN FACTORS

- How to optimize the relationship between people and their activities by the systematic application of the human sciences, integrated within the framework of systems engineering.

TWIN OBJECTIVES OF HUMAN FACTORS

1. Effectiveness of the system. (Safety & Efficiency)

2. The well-being of the individual.

"ERGONOMICS"

- The study of man in his working environment.

HUMAN FACTORS
OTHER THAN
PHYSIOLOGICAL

- Human behavior and performance (Psychological)

    - Decision making and the cognitive processes

    - Design of controls and displays

    - Flight deck and cabin layout

    - Communication

    - Software aspects of computers, maps, charts, and documentation

    - Refinement of staff selection, training, and checking

MULTI-DISCIPLINARY SIDES
OF
HUMAN FACTORS

a. Psychology

b. Physiology

c. Anthropometry and biomechanics

d. Biology and chronobiology

e. Genetics

f. Statistics

HUMAN FACTORS IN AVIATION SAFETY
(SF 320)
LECTURE NOTES

DAY 03   The SHEL Conceptual Model

PO: 1.   *Describe the SHEL conceptual model of human factors in terms of what each of the letters means.*

1.   What do each of the letters in the SHEL conceptual model of human factors represent?

(HF p. 22)

S =   Software

H =   Hardware

E =   Environment

L =   Liveware

2.   What is the most valuable and flexible component in the system?

(HF p. 22)

-    Liveware   - (the human)

-    The central or hub of the SHEL model

PO: 2.   *Describe with examples the various characteristics of the **LIVEWARE** component of the system.*

PO: 3.   *Identify the scientific discipline which is associated with each characteristic of the liveware component.*

3.   What are the various characteristics of the liveware component to consider in the Human Factors conceptual model?

(HF pp. 22-23)

a.   <u>Physical size and shape</u>

-    The human dimensions must be considered early in design

-    Body measurements and movement play a vital role

-    Scientific discipline = Anthropometry and Biomechanics

b. <u>Fuel requirements</u>

- Food - Water - Oxygen

- Scientific discipline = Physiology and Biology

c. <u>Input characteristics</u>   (HF p. 22)

- Sensing to receive light, sound, smell, taste, movement, touch, heat and cold

- Information needed to enable the human to respond to external events

- Scientific discipline = Physiology and Biology

d. <u>Information processing</u>   (HF p. 23)

- The effectiveness of how the sensory information is processed

- Concerned with short and long term memory

- Effected by motivation and stress

- Scientific discipline = Psychology

e. <u>Output characteristics</u>

- Messages sent to and feedback received from the muscles

- The kind and direction of forces needed for the movement of the controls

- Speech characteristics

- Scientific discipline = Biomechanics and Physiology

f. <u>Environmental tolerances</u>

- A narrow range of environmental tolerance

- Concerned with temperature, pressure, humidity, noise, time-of-day, light and darkness

- Individual phobias

- Scientific discipline = Physiology, Biology and Psychology

g.   <u>Individual differences</u>   (HF p. 23)

- People are different

- Some variability around the normal must be anticipated

- Controlled by selection, training, and application of standardized procedures

- Scientific discipline = Psychology

PO: 4.   Describe each of the interfaces of the SHEL human factors conceptual model and give some examples of considerations which apply to each of them.

4.   What must be taken into account in order to make more effective **LIVEWARE-HARDWARE** (L-H) interface in the SHEL conceptual model?

(HF p. 24)

a.   The L-H interface is most commonly considered when speaking of man-machine systems

b.   Designing seats to fit the sitting characteristics of the human body

c.   Designing displays to match the information processing characteristics of man

5.   How does the **LIVEWARE-SOFTWARE** (L-S) interface in the SHEL conceptual model work?   (HF p. 24)

a.   Encompasses the non physical aspects of the system

b.   Includes procedures, manual and checklist layout, symbology and computer programs

c.   Problems are less tangible than L-H and more difficult to resolve

d.   The symbology and conceptual aspects of Head-Up Display (HUD) is an example

6.   Which interface in the SHEL conceptual model was one of the earliest recognized in the flying environment?   (HF p. 25)

- The **LIVEWARE-ENVIRONMENT** (L-E) interface

7.   What are some of the examples where the L-E interface is trying to adapt the human to fit the environment?

(HF p. 25)

-       Fitting flyers with:
    -       helmets against noise
    -       flying suits against cold
    -       goggles against the airstream
    -       oxygen masks against the effects of altitude
    -       Anti-G suits against acceleration forces

8.   What are some of the ways the L-E interface is trying to adapt the environment to fit the human?

(HF p. 25)

-       Pressurization and air conditioning systems and soundproofing

9.   What other L-E consideration is generated in human factors with the speed of transmeridian travel and payload flying 24 hours per day?

(HF p. 25)

-       Problems associated with disturbed biological rhythms and sleep deprivation

-       Growth phenomena   -   teeming passenger terminals

                                    -       increasingly crowded skys

10.   What are the considerations of the **LIVEWARE-LIVEWARE** (L-L) interface of the SHEL conceptual model of human factors?

(HF p. 25)

a.   The team-work of the flightcrew

b.   Interaction between the passengers

c.   Leadership, crew cooperation, teamwork and personality interactions

d.   Instructor/student relationships

e.   Staff/management relationships

(next page)

11. What does human factors attempt to research and explain? (HF p. 26)

- the nature of human behavior and human performance using the human sciences

- it tries to predict how a person will react and respond in a given set of circumstances

SHEL CONCEPTUAL MODEL

S =   Software

H =   Hardware

E =   Environment

L =   Liveware

CHARACTERISTICS OF LIVEWARE

a. <u>Physical size and shape</u>

Anthropometry and Biomechanics

b. <u>Fuel requirements</u>

Physiology and Biology

c. <u>Input characteristics</u>

Physiology and Biology

d. <u>Information processing</u>

Psychology

e. <u>Output characteristics</u>

Biomechanics and Physiology

f. <u>Environmental tolerances</u>

Physiology, Biology and Psychology

g. <u>Individual differences</u>

Psychology

DAY 04   The Nature of Error

PO: 1.   *Be able to describe the nature of human errors in terms of normal distribution and other factors which may affect the distribution of accidents.*

1.   What insight did the Roman orator Cicero give into the nature of human error?

(HF p. 30)

   a.   "It is the nature of man to err."

   b.   "Only a fool preservers in error."

2.   What kind of human error may result from a person reacting in a perfectly normally manner to the situation presented?

(HF p. 30)

   -       Errors that are induced by poorly designed equipment or procedures

3.   What kind of characteristics do errors like these have?

(HF p. 30)

   a.   They are likely to be repeated

   b.   They are largely predictable

   c.   We must look to the designer for correction not the operator

4.   What things did the Three-Mile Island nuclear incident, the Chernobyl nuclear disaster, and the Tenerife double 747 crash have in common?

(HF p. 30)

   a.   They came from human error occurring in a working environment deficient in human factors

   b.   Serious warnings of inadequate human factors and possible devastating consequences were given before the accidents

   c.   Effective measures to heed the warnings were not taken

5.   What are the three basic tenets with respect to human error?

(HF p. 31)

    a.   The origins of error can be fundamentally different

    b.   Anyone can and will make errors

    c.   The consequences of similar errors can be quite different

6.   What does the author consider to be one of the most enduring myths in aviation?

(HF p. 31)

-    That embodied in the term "pilot error"

7.   How are pilot errors different from any other human error?

(HF p. 31)

-    They are, in principle, no different from those made by anyone else

8.   What fallacies are implied by use of the term "pilot error"?

(HF p. 31)

    a.   Somehow the nature of errors made by this operator is unique

    b.   Once the accident cause is attributed to pilot error, the   problem is solved

    c.   The concept focuses more on <u>what</u> happened rather than on <u>why</u> it happened

    d.   The idea does not lead to mishap prevention activity

    f.   It has impeded a more profound and rational examination of human performance

    g.   It has obstructed progress toward greater flight safety

9.   If human error is normal, of what value is the human operator?

(HF p. 31)

    a.   The human is a very flexible system component

    b.   If ergonomics has been properly applied to system design, the human can give increased overall system reliability

10. If human error is considered to be a part of norman human behavior, what question must be answered to plan accurately for man in the system?

(HF p. 31)

- What error rates are considered to be "normal"?

11. What are some normal human error rates?

(HF pp. 31-32)

a. Simple, repetitive tasks are 1 in 100

b. 1 in 1000 is good in most circumstances

c. The rate varies widely depending on factors such as:

- Nature of the task or risk involved
- Motivation
- Sleep loss and fatigue

d. British standards for their automatic landing systems allows for a catastrophic failure of only 1 in 10 million

PO: 2. Describe the term "accident proneness" and its usefulness in mishap prevention efforts.

12. How would you define the term "accident proneness"?

(HF p. 32)

- The tendency of some people to have more accidents than others with equivalent risk exposure, for reasons beyond chance alone

13. How can random distribution account for some people having more accidents than others?

(HF p. 33)

- With 100 accidents amongst 100 people the normal distribution would break down as follows:

Number of Accidents	Number of People
0	37
1	37
2	18
3	6
4	2

14. What might be reasons, other than normal distribution, which could account for some people having more accidents than normal? (HF p. 33)

 a. A greater risk exposure

 b. Inherent awkwardness or lack of muscle coordination

 c. Simple carelessness

15. How do we reduce inherent awkwardness and lack of muscle coordination as reasons for greater than normal accident distribution?

(HF p. 33)

 - With physiological, psychological, and personality deficiency screening

16. What seems to have the greatest effect of the short term factor of simple carelessness? (HF p. 34)

 - Changes in motivation

17. What are the short term influences which may cause a person to have more potential for mishap, and which of them presents the most promise for a fruitful approach to prevention?

(HF p. 33)

Short Term Influences	Most Fruitful for Prevention
a. Changes in motivation	1- Most easily managed
b. Domestic or work induced stress	2- Stress managment
c. Job satisfaction or boredom	3- Possible to manage
d. Health variations	4- Very difficult to manage

18. How does the short term influence of stress affect the human? (HF p. 34)

 - It varies from one person to another

CICERO'S THEOREMS

"It is the nature of man to err."

"Only a fool perservers in error."

ERRORS OF DESIGN

- Are likely to be repeated

- Are largely predictable

- We must look to the designer for correction not the operator

"PILOT ERROR" FALLACIES

- The nature of errors made by this operator is unique

- Once the accident cause is attributed to pilot error, the problem is solved

- The concept focuses more on <u>what</u> happened rather than on <u>why</u> it happened

- The idea does not lead to mishap prevention activity

- It has impeded a more profound and rational examination of human performance

- It has obstructed progress toward greater flight safety

FACTORS FOR VARIABLES
IN HUMAN ERROR RATES

1. Normal distribution

2. Nature of the task (risk involved)

3. Inherent awkwardness or lack of muscle coordination

4. Simple carelessness (lack of motivation)

5. Short term factors

   - Stress

   - Sleep loss and fatigue

NORMAL ERROR DISTRIBUTION

- 100 accidents among 100 people

Number of Accidents	Number of People
0	37
1	37
2	18
3	6
4	2

HUMAN FACTORS IN AVIATION SAFETY
(SF 320)
LECTURE NOTES

DAY 05   Sources of Error

PO: 1.   Explain and give examples of how mismatches between the SHEL components can be a source of human errors.

1.   Using the SHEL model, where are the potential areas of mismatch between the components?

(HF p. 34)

-    At the four interfaces between the components

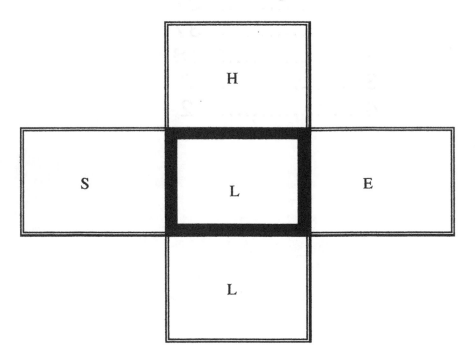

2.   What are some examples of the kind of problems resulting from a poor interface between Liveware-Hardware?

(HF p. 34)

a.   The old three pointer altimeter

b.   Knobs and levers which are badly located or lack proper coding

c.   Poorly designed warning systems

3. What areas of mismatch might come up in the Liveware-Software interface?

(HF p. 34)

a. Deficiencies in conceptual aspects of warning systems

b. Irrational coding system in an operations manual which caused delays and errors in seeking vital information

c. Poorly designed checklists - some without written response

d. A SID with a number that could be confused with the initial departure heading

4. What are some of the conditions which lead to increased errors in the Liveware-Environment interface?

(HF p. 35)

a. Noise, heat, and vibration

b. Reduced performance caused by disturbance of biological rhythms (jet lag)

5. How can deficiencies in the interface between the Liveware-Liveware components develop?

(HF p. 35)

a. Problems with teamwork or crew cooperation

b. An inappropriate authority relationship between the captain and the first officer

c. The optimum "trans-cockpit authority gradient" should be maintained

Use Fig 2.2 ----> (Fig 2.2, p.37)

d. Over 40% of 249 airline pilots surveyed in the UK indicated they had failed more than once to communicate their proper doubts about the operation of the aircraft to the captain to avoid a conflict.

e. Socio-Psychological influences can interfere with proper exchange of information on the flight deck.

f. Socio-cultural influences can also hinder the L-L interface

31

Describe the errors which can occur in the information processing system of the Liveware component in terms of sensing, perception, decision-making, action and feedback.

6.   What are the areas that are the sources of error within the information processing system of the liveware component?

(HF pp. 37-46)

    a.   Sensing

    b.   Perception

    c.   Decision-making

    d.   Action

    e.   Feedback

7.   How do each of these fit together in the human information processing system?

(FIG 2.3, p. 37)

- Use Fig 2.3 ---->

8.   What might be the causes of the individual differences in the sensing process?

(HF p. 37)

    a.   Different types or quality of signals and their background

    b.   Sensitivity differences resulting from:

        -   Physical and mental disorders
        -   Deterioration from age
        -   Interference from lesions or agents such as local   anesthetics

9.   What vital-to-flight information is the human sensory system not designed to detect?

(HF p. 38)

    -   We sense acceleration but not velocity

10.   What is perception?

(HF p. 38)

    a.   The conclusion reached about the nature and meaning of the message received

    b.   The interpretive activity - a major breeding ground for error

11. What are some of the Gestalt laws which are concerned with structure or organizational arrangement of elements?

(HF p. 38)

   a.   Close elements are perceived as a group

   b.   Similar elements are perceived as a group

   c.   Distinguishing the figure form the background
        - Pertinent to visual illusions

12. Which organizational arrangement is the basis for all perception?

(HF p. 38)

   -    Separation of figure and background

13. How does the context of the message affect the way it is perceived?

(HF p. 38)

   -    The stimulus may be the same but the perception of the stimulus is different.

14. What happens to perception when there is ambiguous or inadequate information?

(HF p. 38)

   -    We unconsciously fill in the missing information ourselves.

15. What influence does "set" or expectation have on perception?

(HF p. 38)

   a.   We hear and see what we expect to.

   b.   It is closely related to habit and is particularly applicable to the aviation environment.

16. How does the practice of standardizing equipment and procedures fit into the laws of perception?

(HF p. 39)

   a.   Expectation and habit make it possible for operating habits in one aircraft to transfer to another.

   b.   Design carry-over reduces confusion and minimizes training costs.

17. How can the law of expectation have a detrimental effect on errors?

(HF p. 39)

- It may be difficult to "unlearn" an action that is no longer appropriate.

18. Under what conditions is reversion to an earlier habit pattern more prevalent?

(HF p. 39)

a. When concentration is relaxed

b. Under stress

19. What complicates the relationship of reliability (freedom from errors) to experience?

(HF pp. 39-40)

a. Physiological deterioration begins early and goes on throughout life.

b. Emotional stress often increases with age.

c. Career frustrations only start during the last half of a career.

d. Experience leads to reduction in errors and exposure to situations where errors can have serious consequences.

PO: 3. *Describe how motivation affects performance and what conditions might reduce the motivation level.*

20. What is motivation and what effect does it have on the performance of a particular task?

(HF p. 40)

a. The difference between what a person **can** do and what they **will** do

b. Low motivation = reduced performance with more errors

21.   What factors might bring a person's motivation down?

(HF p. 40)

    a.   Inadequate job satisfaction

    b.   Physical or psychological illness

    c.   Emotional stress or fatigue

    d.   Boredom

PO: 4.   Explain how arousal or alertness relates to how well humans perform.

22.   What is the relationship between the degree of arousal and alertness to the effectiveness of a person in performing a task?

(HF p. 40)

(Refer to Fig 2.4, p. 40)

    a.   A high level of arousal is usually accompanied with a high level of attention.

    b.   High arousal state = high level of activity or

                = preparedness for a high level of mental or physical activity

23.   What happens to performance when arousal is higher or lower than optimum?

(HF p. 41)
    -    Reduced performance is the result

24.   How does the optimum level of arousal for speed differ from that required for accuracy?

(HF p. 41)

    -    The optimum level of arousal for speed is higher than that for accuracy.

PO: 5.   Describe the factors which affect the human decision-making process.

25.   What are the factors which can distort the human decision-making process?

(HF p. 41)
    a.   Emotional or commercial considerations

    b.   Fatigue, medication, motivation

    c.   Physical or psychological disorder

Explain how the false hypothesis or mistaken assumption contributes to error in the human decision-making process.

26. What is the most dangerous characteristic of the false hypothesis?

(HF p. 41)

- It is frequently extremely resistant to correction.

27. What are the situations where the mistaken assumption is more likely to occur?

(HF pp. 41-42)

a. When expectancy is high

b. When attention is diverted elsewhere

c. When it serves as a defense

d. Following a period of high concentration

e. As a result of the effects of motor memory

28. What is the primary drawback to the human decision-making process?

(HF p. 42)

a. Man has a vast capacity for sensory input but only one channel for decision-making.

b. Information must be held in short-term memory until its turn in the decision line.

c. With only one channel it is easily overloaded and load-shedding or filtration can occur.

d. Concentration on a single stimulus may result in ignoring others.

29. How can the action process in the human component be adversely affected?

(HF p. 43)

a. Controls which are badly designed for human factors will contribute to wrong action on the part of the operator.

b. Equipment designed so it can be operated wrong sooner or later will be.

30. How important is feedback to the human information processing system?

(HF p. 43)

a. It is essential to efficient operation.

b. Inadequate or inappropriate feedback can interfere and generate errors.

PO: 7. *Describe some of the conditions which lead to errors in eye witness accounts of accidents.*

31. What are some of the possible influences which can affect the accuracy of eye witness accounts of accidents?

(HF p. 43)

a. Perception

b. Communication factors

c. Retention

d. Circumstances of the event itself

32. What things can impact on the perceptions of the eye witness?

(HF p. 44)

a. Stress

b. Attention selectivity

c. Expectation (Has the greatest significance)

33. What communication factors can be a source of error in the testimony of the eye witness?

(HF p. 44)

a. Suggestion - leading questions

- desire to please the questioner

b. Discussions between a witness and other people - particularly interested parties

c. Trauma of the event if the statement is taken too soon after

d. Threat of criminal charges which may follow

34. How does the issue of retention fit into the accuracy of a witnesses testimony?

(HF p. 45)

    a. Natural decay in memory

    b. Natural tendency to fill in the gaps

    c. An "official" statement will increase future commitment to that position

35. How might the nature of the event influence the testimony of the eye witness?

(HF p. 46)

    a. The trauma of the event may cause distortions, additions or subtractions

    b. The duration and the ambient conditions

36. What precautions should be taken to increase the accuracy of eye witness testimony?

(See Appendix 1.2A pp. 342-343)

INFORMATION PROCESSING ERROR SOURCES

a. Sensing

b. Perception

c. Decision-making

d. Action

e. Feedback

GESTALT LAWS OF PERCEPTION

a. Close elements are perceived as a group

b. Similar elements are perceived as a group

c. Distinguishing the figure form the background

d. Context effects perception

e. Ambiguity will lead to "fill-in-the blank"

f. "Set" or expectation will influence perception

EXPERIENCE
VS
PHYSIOLOGICAL DETERIORATION

a. Physiological deterioration throughout life

b. Emotional stress often increases with age

c. Career frustrations greater with maturity

d. Experience leads to reduction in errors and exposure to situations

MOTIVATION FACTORS

a.   Inadequate job satisfaction

b.   Physical or Psychological illness

c.   Emotional stress or fatigue

d.   Boredom

MISTAKEN ASSUMPTION IS HIGH

a.   When expectancy is high

b.   When attention is diverted elsewhere

c.   When it serves as a defense

d.   Following a period of high concentration

e.   As a result of the effects motor memory

Possible Influences on Eye Witnesses

a. Perception

b. Communication factors

c. Retention

d. Circumstances of the event itself

HUMAN FACTORS IN AVIATION SAFETY
(SF 320)
LECTURE NOTES

DAY 06   Error Classification and Reduction

PO: 1.   *Describe the four ways to classify errors.*

1.   What are the four different ways to classify human errors?

(HF pp. 46-48)

   a.   Design-induced and Operator-induced

   b.   Random, Systematic and Sporadic

   c.   Omission, Commission and Substitution

   d.   Reversible and Irreversible

2.   Which interfaces of the SHEL model are usually involved with design-induced errors?

(HF p. 46)

   a.   L-H or L-S or the interfaces between Liveware and Hardware or Software

   b.   The close proximity of the DC-3 gear and flap handle led to countless cases of confusion between them.

3.   What is the cause of an operator-induced error?

(HF p. 47)

   a.   Inadequate performance on the part of the individual

   b.   Reflecting possible deficient skill, motivation or vision

   c.   Sometimes design-induced error refers to hardware and system-induced error refers to software

4.   What are the differences between the random, systematic and sporadic errors?

(HF p. 47)

   a.   Random errors have no discernable pattern

        -        Many factors may influence the range of variability

b.   Systematic errors are characterized by a small dispersion which is offset from the desired point

        -      Often caused by one or two factors

c.   Sporadic errors occur after routinely good performance

        -      very difficult to predict

5.   How would you define an error of omission?

(HF p. 47)

    a.   Failing to do something which ought to be done

    b.   Missing an item on the checklist

6.   What is an example of an error of commission?

(HF p. 47)

    a.   Doing something which ought not to be done

    b.   Silencing the gear warning horn during configuration

7.   How would you describe an error of substitution?

(HF p. 47)

    a.   Taking action when it is required, but the wrong action

    b.   Shutting down the wrong engine with an engine fire

8.   What is useful about classifying errors as reversible and irreversible?

(HF p. 48)

    -        It is a useful taxonomy when discussing strategies to meet the challenge of human error

9.   What is an example of an error that is reversible?

(HF p. 48)

    a.   An error which allows for a correction

    b.   A computer screen check before entry of the command to   execute

PO: 2.   *Describe some of the differences between human and machine in tasking to reduce errors.*

10.   What is the general principle used to allocate the tasks between human and machine?

(HF p. 48)

-     The system should be designed so the human is given the tasks done best by humans and the machine is given those it does best.

11.   What are the tasks that are best matched to human and machine performance?

(Table 2.1, HF p. 48)

Refer to Table 2.1 ----->

Machine:	Human:
Deductive reasoning	Inductive reasoning
Monitoring	Error correction
Speed	Intelligence
Power	Flexibility & judgement
Consistency	Where there is a need for:
Complex activity	- Emotion
Short term memory	- Motivation
Computation	- Intuition
	- Innovation
Overload or breakdown is sudden	Overload or breakdown is gradual

12.   What did the Royal Air Force (RAF) laboratory tests in 1943 conclude about the human's performance of tasks requiring the monitoring or detection of brief, low intensity and infrequent events over long periods of time?

(HF p. 49)

-     Human performance is poor in those tasks

13. When does the "vigilance decrement" phenomenon usually occur in human performance?

(HF p. 41)

- after about 30 minutes

14. Since practice does not seem to be effective, what is the best way to eliminate the phenomenon of vigilance decrement?

(HF p. 49)

a. Recognize it in the design of equipment and procedures

b. It is relevant to the L-H and L-S interfaces in the SHEL   model.

15. How can the deficiency in human short term memory be reduced?

(HF p. 49)

- Transfer some of the activity to data-link

PO: 3.   *Explain and give examples of the elements in the two-pronged attack on reducing human errors.*

16. What are the two prongs of the two-pronged attack on human errors?

(HF p. 50)

a. Minimizing the occurrence of errors   -   (Probability)

b. Reducing the consequences of errors   -   (Severity)

17. Which of the two prongs seems to have the most relevance in terms of system design?

(HF p. 50)

- The second - Reducing the consequences of errors

18. What basic premise must be accepted to implement the second prong of the attack on human error?

(HF p. 43)

a. Recognition that errors are a part of normal human behavior

b. Dispelling the illusion that it is possible to achieve "error-free" operation

46

19. What does minimizing the occurrence of error focus on?

(HF p. 51)

a. The quality and condition of the L component of the system

b. Ensuring a high level of staff competence through optimizing selection, training and checking

c. Personality, attitudes and motivation play a vital role

d. Tolerance to fatigue and other stresses are important

20. What does minimizing the occurrence of errors mean in the context of the SHEL model?

(HF p. 51)

a. We must work with people as they are rather than what we would like them to be.

b. Interfacing the components of the system with the human as is, with all the rough edges

21. What are some examples of the SHEL interfaces to consider in minimizing the occurrence of error?

(HF pp. 51-52)

a. L-H - Coding for controls and their movement should meet human expectation
  - Displays should not only present information but do it in a way which facilitates the human information processing task

b. L-S - Careful design of the non-physical aspects such as:
  - Checklists
  - Procedures
  - Manuals
  - Charts
  - Airport and route guides

c. L-E - Improve control of noise, vibration and temperature

  (heat has a more adverse effect on mental performance than cold)

d. L-L - Orient training towards enhancing the cooperation and communication between crew members

47

22. In the SHEL context, what matching is required to minimize the occurrence of errors?

(HF p. 52)

- Optimum matching of the components of the system with the characteristics of the human

23. How does designing for minimizing the occurrence of errors fit in with the level of arousal?

(HF p. 52)

Refer to Fig 2.4, HF p. 40 --->

a. Aircraft design and operating procedures must aim to establish an optimum level of arousal.

b. Optimum level of arousal depends on the nature of the task to be performed

c. A complex task requires less arousal than a simple task

24. How does the classification of errors as random, systematic or sporadic fit into the attack on minimizing the occurrence of errors?

(HF p. 52)

a. Random - A variety of origins

- Personnel selection, training, checking and supervision may all play a role

b. Systematic - Usually caused by only one or two factors

- Normally easier to correct once a proper analysis is made

- A good operator feedback system will help

c. Sporadic - Difficult to predict or reduce

- Are usually resistant to correction through training or indoctrination

25. What are some of the principles connected with the second prong of reducing the consequences of errors?

(HF pp. 53-54)

a. Design equipment so that errors are reversible

- Visual display units of computer and navigation systems use a "scratch pad" which can be checked before entry into the computer.

b. Developing a thorough and efficient cross-monitoring

- Redundancy in flight crew employs this concept

- CRM is the employment of cross-monitoring

- Equipment monitoring of human performance:

- Ground proximity warning system
- Altitude alerting system

ERROR CLASSIFICATIONS

1. Design-induced and Operator-induced

2. Random, Systematic and Sporadic

3. Omission, Commission and Substitution

4. Reversible and Irreversible

ERROR CLASSIFICATIONS (CONT)

3. Omission, Commission and Substitution

4. Reversible and Irreversible

HUMAN OR MACHINE EFFICIENCY

Machine:	Human:
Deductive reasoning	Inductive reasoning
Monitoring	Error correction
Speed	Intelligence
Power	Flexibility & judgement
Consistency for:	Where there is a need
Complex activity	- Emotion
Short term memory	- Motivation
Computation	- Intuition
	- Innovation
Overload or breakdown is sudden	Overload or breakdown is gradual

THE TWO-PRONGED ATTACK
ON
HUMAN ERROR

1. Minimizing the occurrence of errors

    (Probability)

2. Reducing the consequences of errors

    (Severity)

MINIMIZING THE OCCURRENCE OF ERRORS

a. Random   - A variety of origins

                    - Personnel selection, training, checking and supervision

b. Systematic   - Usually caused by only one or two factors

                    - Proper analysis

                    - Operator feedback system

c. Sporadic   - Difficult to predict or reduce

                    - Resistant to correction

HUMAN FACTORS IN AVIATION SAFETY
(SF 320)
LECTURE NOTES

DAY 07   Fatigue, Body Rhythms, and Sleep

PO: 1.   *Describe the causes of jet lag and fatigue and how these phenomena effect human performance.*

1.   How would you define the term "Jet Lag"?

(HF p. 56)

-        Lack of well-being experienced after long distance air travel

2.   What are the more specific performance symptoms of jet lag?

(HF p. 56)

-        Performance, motivation, mood and behavior suffer to some degree

3.   What are the dangers of flying while under the influence of jet lag?

(HF p. 56)

-        Slowed reaction and decision time

-        Defective memory for recent events

-        Errors in computations

-        Tendency to accept lower standards of operational performance

-        Sleep disturbance and deprivation

-        Crew sleeping at duty station

4.   How prevalent in the air carrier industry is it for flight crews to be fighting the effects of jet lag?

(HF p. 57)

-        It is a universal symptom of long range air carrier operations.

5. How does the Galvanic Skin Response (GSR) system assist in countering some of the effects of jet lag?

    (HF p. 57)

    -   Monitors skin resistance to detect low arousal

6. How big a problem is fatigue in the aviation industry?

    (HF p. 58)

    a. 93% of pilots report it as a problem

    b. 85% of pilots said they felt extremely tired or "washed out" in the preceding 30 days

    c. NASA reported, in 1981, that fatigue associated decrements in performance resulted in substantive potentially unsafe aviation conditions

    d. British CHIRP system announced, in 1984, that the largest number of reports received concerned fatigue, sleep and the way the work patterns are constructed

    e. Challenger space shuttle disaster in 1986 considered contributing human factors of:

        -   Sleep loss, excessive duty shifts, circadian or daily rhythm effects and the resulting fatigue on the decision to launch the shuttle in spite of concern about safety

7. What indications are there for greater formal education in the human factors associated with fatigue?

    (HF p. 58)

    -   Some reports show depressing signs of ignorance and complacency

8. What are the four causes of fatigue?

    (HF p. 57)

    1- Inadequate rest

    2- Disruption or displaced body rhythms

    3- Excessive muscular or physical activity

    4- Excessive cognitive activity

9.   What are some other conditions that can contribute to fatigue?

(HF pp. 58-59)

   a.   Low humidity at high altitude and noise

      -   3% compared to the normal 40-60%

   b.   Excessive and frustrating customs and security procedures

PO: 2.   *Describe some of the body rhythms and how they relate to human factors of performance.*

10.   Which of the body rhythms is the most significant?

(HF p. 59)

   a.   The most significant is the 24 hour

   b.   This relates to the 24 hour rotation of our planet

11.   When did the first chronobiological reports in the literature appear?

(HF p. 59)

      -   First scientific reports appeared more than 200 years ago

12.   What serves as the focal point of scientific study in chronobiology?

(HF p. 59)

      -   An International Society of Chronobiology

13.   What are some of the cycles of body chemistry which coincide with the 24 hour oscillations?

(HF p. 60)

   a.   Sodium and potassium excretion

   b.   Amino acids, cortisol and other hormone levels

14.   What does the term acrophase have reference to in chronobiology?

(HF p. 60)

      -   The highest point in the rhythm curve for each of the body's systems

15. Which single circadian pacemaker or controlling mechanism governs the body's biorhythm?

(HF p. 60)

   a. There is no single pacemaker

   b. Substantial control is in the suprachiasmatic nuclei of the hypothalamus deep inside of the brain

   g. The free running cycle ranges from 24 to 27 hours

16. What are some of the rhythm and time cues that maintain the 24 hour cycle, and what are they are called?

(HF p. 60)

   - *zeitgebers* or entraining agents include:

      - Meals
      - Physical and social activity
      - Sleep

17. How does the oral temperature rhythm indicate the daily cycle?

   Refer to Fig 3.1 ---->   (HF p. 61)

   a. The peak temperature is reached during the evening

   b. Extroverts and evening types tend to peak later than introverts and morning types

18. What does the Birthdate Biorhythm Theory have to offer?

(HF p. 61)

   a. Consists of three long term cycles:

      1- Physical   = 23 days

      2- Emotional   = 28 days

      3- Intellectual   = 33 days

   b. Statistical analysis of a very large number of aircraft accidents has failed to establish any correlation between the three cycles and the timing of aircraft accidents

19. What can be said about the rhythm of performance in the 24 hour cycle?

Refer to Fig 3.2 ----> (HF p. 62)

a. The curve is task-dependent

b. Will vary according to the task

c. Maximum and minimum performance scores within a cycle is task dependent

 - tends to be greater with complex than simple tasks

c. Reduction of performance during certain parts of the 24 hour period is not a result of sleep deprivation

d. Practice, motivation and increased effort will raise and flatten the curve

e. Loss of performance from the natural cycle may be greater than that coming from loss of a single night's sleep

f. There is a post-lunch dip not shown on Fig 3.1

 - May want to avoid critical tasks which require optimum performance

20. What kind of an influence does the disturbance of biological rhythms have in aviation? (HF p. 63)

a. It is a pattern of life for the long-range flight crew

b. Results from irregular work schedules superimposed upon time zone changes

21. What are some of the terms associated with the disturbance of biological rhythms? (HF p. 63)

a. Circadian dysrhythmia or desynchronosis or the pilot preferred term "transmeridian dyschronism"

b. Called metergic dyschronisim in shift workers

22. What are the four factors which are concerned with adapting to living environment changes?

(HF p. 64-65)

    a.   Systems shift their phase at different rates

        -    Out of phase with local time

        -    Out of phase with each other

        Refer to Table 3.1 ---->   (HF p. 64)

    b.   Resynchronisation occurs at different rates depending on the need to advance (eastbound) or delay (westbound) the phase

        -    Most travelers find recovery from eastbound flights is harder than from westbound flights

    c.   Resynchronisation does not appear to have a constant rate

        Refer to Table 3.2 ---->   (HF p. 65)

    d.   There are substantial differences in the ability of individuals to adjust their circadian rhythms to repeated transmeridian shifts

23. How has the development of chronohygiene drugs progressed?

(HF p. 65)

    a.   There is a demand for a drug which would accelerate the resynchronisation

    b.   Experimental work using various drugs has not yet proved conclusive

24. What things can be used to enhance the chronohygiene adjustment of the phase shift of the biorhythm?

(HF p. 66)

    a.   A program involving timing of meals, use of drugs, exercise, sleep and social cues was proposed to accelerate the phase shift

    b.   A self-imposed living schedule following transmeridian flight may modify the rate of physiological and psychological adjustment to the new time zone

PO: 3.   *Describe the characteristics of the two basic kinds of sleep and how sleep effects the physiological well being of the human.*

25.   What is the most common physiological symptom of long-range flying?

(HF p. 66)

   -      Disturbance of the normal sleep pattern is the most

26.   What causes this disruption in the normal sleep pattern?

(HF p. 66)

   -      Arises from having to work (travel) during the normal sleep period and trying to sleep during a time when biological rhythms are not set for sleeping

27.   How would you explain the difference between "monophasic" and "polyphasic" in describing sleep patterns?   (HF p. 66)

   a.   Monophasic = one long period a day

   b.   Polyphasic = several periods during the day

   c.   Once established, the monophasic pattern becomes the normal rhythm of the brain

28.   What is meant by the "Golden age of sleep"?   (HF p. 66)

   -      Early 1950s to mid-1960s

   -      Much research and learning about the nature of sleep

   -      The precise function of sleep is still obscure

29.   What are the two basic kinds of sleep?   (HF p. 67)

   a.   Two basic kinds;

          -      Orthodox

          -      Paradoxical or REM (Rapid Eye Movement)

   b. Each has its own characteristics

          Refer to Table 3.3 ---->   (HF p. 67)

30. How does the pattern of sleep change with age?

(HP p. 67)

- At birth = 50% Orthodox and 50% REM

- Adulthood = 20% REM

- By age 70 = 15% REM

31. What are the four subdivisions of orthodox sleep?

(HF p. 68)

a. Stage 0 - 4
        Refer to Table 3.4 ----> (HF p. 68)

b. During a normal night, sleep shifts from one stage to another about 30 times

c. REM sleep occurs about once every 90 minutes
        Refer to Fig 3.5 ----> (HF p. 68)

32. What are the three factors affecting the recuperative effect of naps?

(HF p. 69)

a. The time of day the nap is taken

b. Hours of prior wakefulness

c. Nap duration - Varies with individuals

33. How can you apply the benefits of a nap?

(HF p. 69)

a. Before a long night flight or after arrival, to try and recover some of the sleep loss

b. Not less than 10 minutes duration for it to be restorative

34. What is microsleep, when does it usually occur, and what benefit does it bring?

(HF p. 69)

a. Vary from a fraction of a second to up to 2-3 seconds

b. Person is not generally aware of them

c.   Occur more frequently during conditions of fatigue

d.   Not helpful in reducing sleepiness

35.   What can be said about the quality of sleep?

(HF p. 69)

     a.   No simple way to quantify

     b.   Often left to subjective reporting

     c.   Depth of sleep is also hard to quantify

          -      Stage 4 is deep sleep

          -      However arousal from REM is about same as from Stage 4

     d.   Restorative quality is better than depth description

36.   To what does the term "anchor sleep" refer?

(HF p. 69)

     -      Consists of at least four hours of uninterrupted sleep at the individual's home domicile sleep time

37.   How does sleep affect memory?

(HF p. 70)

     a.   There seems to be an increase in information retention just before dropping off to sleep

     b.   Sleep and night-time are better for memory than wakefulness and daytime

     c.   Restorative processes are enhanced

          -      Due to the increase in the net rate of protein synthesis during sleep

PO: 4.   *Differentiate between the effects and causes of clinical and situational insomnia.*

38.   What are the two types of insomnia?

    a.   Clinical insomnia:

    b.   Situational insomnia:

39.   What are types and possible causes for clinical insomnia?

(HF pp. 70-71)

    a.   Difficulty sleeping under normal conditions in phase with body rhythms

    b.   Three types:

        1.   Inability to get to sleep
        2.   Waking and not being able to return to sleep
        3.   Early waking in the morning

    c.   Rarely a disorder itself but usually a symptom of another disorder

40.   What are the causes for situational insomnia?

(HF p. 71)

    a.   Difficulty in sleeping in a particular situation when biological rhythms are disturbed

    b.   A wide difference in individuals and their ability to sleep out of phase with their biological rhythms

    c.   Emotional stress factors may be involved also

    d.   Usually physiological and related to body chemistry

PO: 5.   *Describe how some of the commonly used drugs affect sleep.*

41.   How do various drugs affect sleep?

(HF pp. 71-74)

Barbiturates   (overdose fatal)
     &
Benzodiazepines (Valium):

- Induce sleep

- Seriously addictive

- Adverse affect on performance

- Combined with alcohol is unpredictable

- Can sometimes cause a hangover effect

Alcohol:

- Central nervous depressant

- Induces sleep (Abnormal pattern - suppression of REM sleep)

Caffeine:

- Increases alertness and normally reduces reaction times

- Disturbs sleep reduces stage 4 and REM

Amphetamines:

- Helps maintain performance during sleep deprivation

- Postpones the effect of sleep loss

- Disturbs the following sleep

42.   What is the function of sleep?

(HF p. 74-76)

a.   The answer to the question is still hypothetical

b.   A rebound effect takes place following deprivation of a certain type of sleep, suggesting a need

c.   The first five hours contains most of the slow wave sleep (SWS) and may be the most important

d.   Appears to play a role in the maintenance of motivation

PO: 6.   *Describe the performance characteristics of a sleep-deprived person.*

43.   What are the performance characteristics of the sleep-deprived person?

(HF p. 77-78)

a.   Lapses and inconsistency in performance

b.   Performance decrement increases with altitude and higher workload

c.   Subjective attitudes, appearance, behavior, and mood suffer

-     Even small amounts of loss lower motivation

d.   The more complex task suffers more than the simple one

e.   The more interesting task suffers less than the monotonous or duller one

f.   Most dangerous - the person with sleep-loss induced performance degradation is unlikely to be aware of the deteriorating performance

44.   What operational considerations must be made with regard to sleep loss and human performance?

(HF p. 72)

a.   Sleep deprivation should be avoided

b.   Sleep disturbance should be minimized

c.   An enlightened design and supervision of crew scheduling is needed

d.   Crew sleeping facilities or on duty naps

(HF p. 79)

45.   How effective are drugs and hormones in helping with control of biorhythm?

(HF p. 81)

-     No totally effective and acceptable drug has been developed for routine and long-term application in the control of the problems associated with transmeridian dyschronism.

46. How is Autogenic Training set up to work?

(HF p. 82)

a. Uses passive concentration on certain formula

b. Produces a state of psychophysiological relaxation

c. Facilitates enhancement of certain homeostatic, self-regulating mechanisms in the body

d. Requires skilled teaching - not learn-it-yourself

47. What benefit does Autogenic Training have on improving sleep?

(HF p. 82)

- Improvement in sleep both at home and in the traveling environment

48. How can exercise affect sleep?

(HF p. 83)

a. It increases the slow wave sleep (SWS).

JET LAG

- Lack of well-being experienced after long distance air travel

Dangers:

- Slowed reaction and decision time

- Defective memory for recent events

- Errors in computations

- Tendency to accept lower standards of operational performance

- Sleep disturbance and deprivation

- Crew sleeping at duty station

FOUR CAUSES OF FATIGUE

1- Inadequate rest

2- Disruption or displaced body rhythms

3- Excessive muscular or physical activity

4- Excessive cognitive activity

FACTORS AFFECTING
RECUPERATIVE EFFECT OF NAPS

a.   The time of day the nap is taken

b.   Hours of prior wakefulness

d.   Nap duration

TYPES OF INSOMNIA

a.   Clinical insomnia:

b.   Situational insomnia:

Barbiturates   (overdose fatal)
        &
Benzodiazepines (Valium):

-       Induce sleep
-       Seriously addictive
-       Adverse affect on performance
-       Combined with alcohol is unpredictable
-       Can sometimes cause a hangover effect

Alcohol:

-       Central nervous depressant
-       Induces sleep
        (Abnormal pattern - suppression of REM sleep)

Caffeine:

-       Increases alertness and normally reduces
        reaction times
-       Disturbs sleep reduces stage 4 and REM

Amphetamines:

-       Helps maintain performance during sleep
        deprivation
-       Postpones the effect of sleep loss
-       Disturbs the following sleep

PERFORMANCE CHARACTERISTICS
OF
SLEEP-DEPRIVED PERSON

a.   Lapses and inconsistency in performance

b.   Performance decrement increases with altitude and higher workload

c.   Subjective attitudes, appearance, behavior, and mood suffer

d.   The more complex task suffers more

e.   The more interesting task suffers less

f.   The person unlikely to be aware of the deteriorating performance

AUTOGENIC TRAINING

a. Uses passive concentration on certain formula

b. Produces a state of psychophysiological relaxation

c. Facilitates enhancement of certain homeostatic, self-regulating mechanisms in the body

d. Requires skilled teaching - not learn-it-yourself

HUMAN FACTORS IN AVIATION SAFETY
(SF 320)
LECTURE NOTES

DAY 9   Fitness and Performance

PO: 1.   *Describe total and partial incapacitation, their causes, differences and the operational considerations for countering their effects.*

1.   What are the extreme pathological conditions which can lead to sudden, total incapacitation?   (HF p. 84)

   -      Cardiovascular disorders:

          -      Heart attack   = 1/7 th of incapacitation

          -      Stroke

2.   What is the cause of the more frequent in-flight incapacitation of pilots?
                                                                                         (HF p. 84)

   -      Gastrointestinal disorders:

          -      More frequent   = 1/2 of incapacitation

          -      Usually from contaminated food

          -      Airlines insist the flightcrew eat different meals

          -      Sometimes prohibit the consumption of seafood

3.   What type of operational measures are currently being taken to minimize the risk of sudden incapacitation?   (HF p. 85)

   a.   Operational design of procedures

   b.   Emergency procedures training in the simulator

   c.   Pilots are hesitant to take control from another qualified pilot

4.   From an operational standpoint, what is the difference between total and partial incapacitation?   (HF p. 85)

   a.   Any reduction in capacity

   b.   Partial incapacitation may be far more insidious

   c.   Frequently not apparent to others

   d.   Sometimes not apparent to the one involved

5.   What are some of the reasons for partial incapacitation?   (HF p. 85)

   a.   Fatigue

   b.   Stress

   c.   Sleep and biological rhythm disturbance

   d.   Medication

   e.   Could result from lowered motivation rather than a condition

   f.   Pathological condition

           -      Hypoglycemia or low blood sugar

PO: 2.   Define fitness and explain how physical fitness relates to mental fitness and overall human performance.

6.   How would you define fitness?   (HF p. 86)

   a.   A condition which permits a generally high level of physical and mental performance.

   b.   An ability to perform with minimal fatigue, to be tolerant to stress and to be readily able to cope with changes in the environment.

7.   How important is a pilot's physical condition as it affects performance as a pilot?
                                                                                   (HF p. 86)
        -      Poor condition: -   More subject to error

                                    -      Poor judgement

74

- Good condition: - Mentally alert

- Greater capacity for arduous mental work

8. What is the connection between physical fitness and mental performance?
(HF p. 86)

- There is a Physiological ----> Psychological link

FITNESS   PERFORMANCE

Improved Body Systems   Brain

Improved Self-esteem

Improved Emotional State   Mental

Motivation

Psychological

Reduced Tension   Psychomotor

Resistance Against Fatigue   Physiological

Physical

- Physical fitness ----> Mental health   -   Less depression

-    Less tension and anxiety

-    Improved self esteem

-    More motivation

- Physical fitness -----> Physical health   -   Less sickness

-    Fatigue resistance

-    Less fatigue

-    Recover faster

PO: 3. *Describe how each of the six main factors which affect fitness, also effect human performance in the flying environment.*

9.   What are the main factors which affect fitness and performance?

POSITIVE	NEGATIVE
Exercise	Smoking
	Alcohol
	Drugs
Stress Management	Stress
Proper Diet	

10.   What are the three general benefits, attributed to fitness, achieved by a suitable exercise program?   (HF pp. 87-88)

a.   Feels better

b.   Looks better

c.   Is healthier

11.   How does a person who feels better from an exercise program benefit?

(HF p. 87)

a.   Improves moral and motivation

b.   Less irritability

c.   Improvement in postural tone of skeletal muscles

d.   Improves appetite - enjoy food more

12.   What does looking better from a good exercise program do to benefit personal performance?   (HF p. 88)

a.   Reinforces feeling better

b.   Enhanced motivation and thus performance

76

13.   In what ways is a person with a good exercise program actually healthier?

(HF pp. 87-88)

    a.   Keeps weight normal and thus reduces potential for:

-     Arthritis

-     Back pain

-     Breathing problems

-     Cardiovascular disorders

    b.   Improves heart condition:

-     Increased heart stroke volume

-     Lower heart rate

-     Better heart efficiency

-     Better extraction of oxygen from arterial blood

-     More effective redistribution of blood to the working muscles

-     Improved peripheral blood circulation regulation

    c.   Reduces chance of heart attacks and high blood pressure

    d.   Greater tolerance to heat or cold

    e.   Exercise tends to dissolve tension and stress

14.   What are the three types of exercises which should be included in the effective physical conditioning program?

(HF p. 88)

    a.   Mobility

-     To ensure all the major joints and muscles move freely throughout their full range

    b.   Strengthening

-     To build some cushion of reserve for special occasions

c. Cardiovascular (Heart/Lung) (Aerobics)

  - To improve condition of the heart muscle

  - To increase the amount of oxygen the body can process

15. What is the best way to achieve this three fold kind of fitness?

(HF pp. 88-89)

a. Rarely will normal work give what is needed

b. Need for a systematically designed and medically approved program

c. The program should be enjoyable

16. How do some of the usual "exercise" activities provide for fitness?

(HF pp. 89-90)

a. Walking/Jogging:

  - Walking is a good introduction to fitness

  - Need for a medical check within a few months before starting a fitness program

  - Jogging - heart rate should be monitored occasionally during the vigorous exercise:

200	200
- Age	-50
	150
- Unfitness handicap (40 to 0)	-10
	140

  - Heart rate should return to below 100 after 10 minutes

b. Cycling:

  - Has to be vigorous enough to keep an elevated heart rate

c. Swimming:

  - Excellent form of exercise

  - Has to be vigorous enough to keep an elevated heart rate

78

d. Ball games:

  - Generally less effective

  - Golf and cricket have no significant aerobic effect

e. Hatha Yoga:

  - May be some increase in stress tolerance

17. What are the three significant components of tobacco smoke which destroy fitness?

(HF p. 90)

  a. Nicotine

  b. Tar

  c. Carbon monoxide

18. What are the main health hazards statistically linked to smoking?

(HF p. 91)

  a. Cancer

  b. Cardiovascular disease

19. What effect does nicotine have on the body?

(HF p. 91)

  a. The source of satisfaction and addiction in smoking

  b. Increases adrenaline and non adrenaline output

  c. A raised level of physiological arousal

  d. Smoker's reaction times increase when deprived of tobacco

  e. Complex task performance is worse

  f. Short-term memory somewhat worse

  g. Long-term memory somewhat better

20. What effect does tar have on the body?

(HF p. 90)

- Tar in tobacco smoke is a carcinogen (causes cancer)

21. How does carbon monoxide effect health and fitness of the body?

(HF p. 92-93)

a. Haemoglobin is attracted to carbon monoxide 210 times more than it is to oxygen

b. Results in an oxygen deficiency to the brain

c. Aerobic performance of smokers is significantly worse than non-smokers

- Raises the physiological altitude (5,000 = 10,000)

d. Deteriorates central nervous system (CNS) function:

- Visual discrimination

- Judgment

- Manual dexterity

- Memory

- Vigilance task performance

- Psychomotor function

22. What is the estimated cost of alcoholism in industry and the overall cost of alcohol abuse in the U.S. annually?

(HF p. 95)

a. Alcoholism in industry = $ 10 billion / Year

b. Overall alcohol abuse = $ 25 billion / Year

23. How does alcohol affect performance?

(HF p. 95)

a. Impairs discrimination

b.   Impairs visual and auditory perception

c.   Disrupts short-term and long-term memory

d.   Impairs Thinking and decision making

e.   Impairs coordinated hand-eye movements

f.   Slows reaction time

g.   Lowers inhibitions and increases recklessness

h.   Risk taking increases

i.   Errors in judgement pass unnoticed

24.   What are the insidious flying performance implications for alcohol users which may not be generally known?

(HF p. 95)

a.   Higher mental and reflex functions can be affected for two to three days after a "serious drinking session"?

b.   Altitude increases the performance-degrading effect

25.   What is the accident rate for those who abuse alcohol compared to other workers?

(HF p. 96)

-      Three times higher

26.   How does the blood alcohol content (BAC) relate to flying performance of a pilot?

-- Fig 4.3 ------>   (HF p. 97)

a.   Safety Pilot Intervention

       BAC

       40 mg%   = 1

       80 mg%   = 3

       120 mg%   = 16

b.   Hangover effects can increase flight planning errors seven times

27.   What part does alcohol play in the aviation accident record?

(HF p. 97)

-       About 20% of general aviation pilots killed in the U.S. were found to have a
        BAC of 15 mg% or higher

28.   What are some of the indicators a friend or supervisor might observe in a pilot who is
        tending toward alcoholism?   (HF p. 98)

                    See Append. 1.9 ---->   (HF p. 344)

29.   What part do on the job use of drugs play in aviation safety?

(HF p. 98)

        a.   On the job use of drugs by pilots

                Frequent or occasional   = 25%

                Rarely ....   = 21%

                Total ....   = 46%

        b.   Nearly all known drugs have an adverse effect on the performance of skilled
                tasks.

30.   How prevalent are drugs in our society?

(HF p. 98)

TYPE OF DRUG	DIFFERENT BRANDS OR NAMES
Basic prescription only	= 7,000
Over-the-Counter (U.S.)	= 300,000

31.   What impact does drugs and alcohol have on the aviation accident record in recent
        years?

                From: *The Federal Air Surgeon's Medical Bulletin*
                            Spring 1992

        a.   About 26% of all fatal aviation accidents have drug or alcohol involvement.

b.   Both alcohol and over-the-counter drugs are involved in 8% for each.

c.   Controlled substances and prescription drugs account for a combined total of 10%.

32.   What is a "Stressor"?   (HF p. 99)

-      An event or situation which induces stress

33.   What are some of the adverse results of stress in the human?

                                                                                          (HF p. 99)

a.   Job dissatisfaction

b.   Reduced work effectiveness

c.   Behavior changes

d.   Health damage

e.   Human system breakdown

34.   What are the **physiological** stressors the typical pilot has to face?

                                                                                          (HF p. 99)

a.   Noise

b.   Vibration

c.   Temperature extremes

d.   Humidity extremes

f.   Acceleration forces

g.   Circadian rhythm disruption

35.   What are some of the **psychological** stressors a pilot has to deal with?

                                                                                          (HF p. 100)

a.   Feelings of insecurity

-      Six month medical and proficiency checks

b.   Job security

c.   Role overload

d.   Emotional stress

    -    Family separation

e.   Domestic stress

36.   How does the Life Change Units (LCUs) system of measuring potential stress work?

(HF p. 101)

Show Append. 1.10 ------>   (HF p. 344)

a.   Total number of LCUs are cumulative

b.   The risk of becoming involved in an accident increases as the total increases.

37.   What are the two most commonly applied means of managing stress?

(HF p. 101)

a.   Drugs

b.   Alcohol

38.   Where does the damage from the stressor come?

(HF p. 101)

-    From the individual's response to it.

-    Not from the stressor itself

(next page)

39. What are some of the more effective means of managing the response individuals make to the stresses in their lives?

(HF pp. 101-102)

a. Autogenic Training

- Non-specific self-relaxation

b. Advisory groups

c. Employee Assistance Programs (EAP)

d. Comprehensive Program

40. How does the "Comprehensive Program" at work assist individuals in managing or reducing the stress in their lives?

(HF p. 103)

- Show Fig 4.4 ----->

(HF p. 103)

41. What is the keystone to the comprehensive program of stress management?

(HF p. 103)

- The test which provides for each individual an analysis of their own psychological and biochemical stress profile

42. What is the primary function of food in the diet?

(HF p. 104)

- To provide energy to build and repair the tissues of the body

43. What are the seven types of intake the human body needs to function effectively?

(HF pp. 104-106)

a. Carbohydrates:

- The chief and most immediate source of energy

(next page)

b. Fats:

- The most concentrated source of heat energy

- Can be stored in larger quantity than any other food

- Take longer to digest

- Contain twice the calories as carbohydrates

c. Proteins:

- Needed for building and repair of body tissues

- A more lasting source of nourishment than carbohydrates

- Vegetable may be preferable to animal

- Raw is preferable to cooked

e. Minerals:

- Have no energy producing value

- Essential to maintaining health

f. Water:

- Prevents dehydration

g. Fiber:

- Essential for proper functioning of the digestive and bowel process

- Comes mainly from plants (Fresh fruits and vegetables)

- Low fiber diet contributes to constipation, hemorrhoids, diabetes and certain forms of cancer

(next page)

h.   Vitamins:

-    Body requires in very small amounts

-    No food contains the entire range needed

-    Excess of water-soluble is probably no problem

-    Excess of fat-soluble can be dangerous

-    Major sources in Append 1.11 ----->   (HF p. 345)

44.   What are the three main types of food to provide the body with a balanced diet?
                                                               (HF pp. 105-106)

a.   Nutrients:

-    Proteins, fats, carbohydrates, vitamins, minerals and water

b.   Energy:

-    calories

c.   Dietary fiber:

-    A complex mixture of plant substances

SUDDEN TOTAL PILOT INCAPACITATION

- Cardiovascular disorders:

    - Heart attack

    - Stroke

- Gastrointestinal disorders:

PARTIAL INCAPACITATION

a. Fatigue

b. Stress

c. Sleep and biological rhythm disturbance

d. Medication

e. Pathological condition

f. Lowered motivation

FITNESS

- A condition which permits a generally high level of physical and mental performance.

- An ability to perform with minimal fatigue, to be tolerant to stress and to be readily able to cope with changes in the environment.

PHYSICAL FITNESS = HEALTH

- Physical fitness ----> Mental health

  - Less depression

  - Less tension and anxiety

  - Improved self esteem

  - More motivation

- Physical fitness -----> Physical health

  - Less sickness

  - Fatigue resistance

    - Less fatigue

    - Recover faster

FACTORS EFFECTING PERFORMANCE

POSITIVE   NEGATIVE

Exercise   Smoking

                                                  Alcohol

                                                  Drugs

Stress Management   Stress

Proper Diet

COMPONENTS OF TOBACCO

WHICH DESTROY FITNESS

a. Nicotine

b. Tar

c. Carbon monoxide

HEALTH HAZARDS OF SMOKING

a. Cancer

b. Cardiovascular disease

EFFECTS OF NICOTINE

a. Addiction

b. Increases adrenaline and non adrenaline output

c. A raised level of physiological arousal

d. Smoker's reaction times increase when deprived of tobacco

e. Complex task performance is worse

f. Short-term memory somewhat worse

g. Long-term memory somewhat better

CARBON MONOXIDE EFFECTS

a.   Oxygen deficiency to the brain

b.   Significantly worse aerobic performance

c.   Deteriorates central nervous system (CNS)
     function:

- Visual discrimination

- Judgment

- Manual dexterity

- Memory

- Vigilance task performance

- Psychomotor function

ANNUAL ALCOHOL COSTS

Alcoholism in industry   = $ 10 billion

Overall alcohol abuse   = $ 25 billion

ALCOHOL AND PERFORMANCE

a.   Impairs discrimination

b.   Impairs visual and auditory perception

c.   Disrupts short-term and long-term memory

d.   Impairs Thinking and decision making

e.   Impairs coordinated hand-eye movements

f.   Slows reaction time

g.   Lowers inhibitions and increases recklessness

h.   Risk taking increases

i.   Errors in judgement pass unnoticed

SAFETY PILOT INTERVENTIONS
WITH
ALCOHOL IMPARED PERFORMANCE

BAC	INTERVENTIONS
40 mg%	= 1
80 mg%	= 3
120 mg%	= 16

ON THE JOB PILOT DRUG USE

Frequent or occasional	= 25%
Rarely	= 21%
Total	= 46%

How prevalent are the drugs in our society?

TYPE OF DRUG	DIFFERENT BRANDS OR NAMES
Basic prescription only	= 7,000
Over-the-Counter (U.S.)	= 300,000

ADVERSE RESULTS OF STRESS

a. Job dissatisfaction

b. Reduced work effectiveness

c. Behavior changes

d. Health damage

e. Human system breakdown

PHYSIOLOGICAL PILOT STRESSORS

a. Noise

b. Vibration

c. Temperature extremes

d. Humidity extremes

f. Acceleration forces

g. Circadian rhythm disruption

PSYCHOLOGICAL PILOT STRESSORS

a. Feelings of insecurity

b. Job security

c. Role overload

d. Emotional stress

e. Domestic stress

Results of the Life Change Scale

Below 150   -   Little or no problem

150 - 199   -   Mild problem

-    A 37% chance you'll feel the
     impact of stress with physical
     symptoms

200 - 299   -   Moderate problem

-    A 51% chance of experiencing a
     stress related illness or accident

300 & over   -   DANGER!

-    Stress is threatening your well
     being.

-    An 80% of a stress related illness
     or accident

DIETARY INTAKE REQUIREMENTS

1. Carbohydrates:

2. Fats:

3. Proteins:

4. Minerals:

5. Water:

6. Fiber:

7. Vitamins:

THREE MAIN TYPES OF FOOD

a.   Nutrients:

b.   Energy:

c.   Dietary fiber:

DAY 10   Vision

PO: 1.   *Define the various terms associated with the measurement of light.*

1.   What is the measurement of light called?

(HF p. 107)

-   **Photometry**

2.   What are the three basic terms used to determine if the lighting is sufficient?

Refer to Table 5.1 ---->   (HF p. 107)

a.   **Intensity** =   Point source brightness

b.   **Illumination** =   Light falling on the surface

c.   **Luminance** =   Brightness of the reflected light from the surface

3.   To what does the term **contrast** refer?

(HF p. 108)

-   The relationship of brightness or luminance of a target to its background or surroundings

4.   What is meant by **refraction**?

(HF p. 108)

-   The bending of light as it goes through a medium of different density

5.   What causes the eye to see **color**?

(HF p. 108)

-   The different wavelengths of the light entering the eye resolve as different colors

PO: 2.　　　*Identify the various terms associated with the physical makeup of the eye and describe their function.*

6.　　What is the function of each of the following parts of the eye?

(HF p. 109-110)

-	**Pupil**	-	Controls the amount of light entering the eye
-	**Iris**	-	Gives color to the eye governs the size of the pupil
-	**Ciliary muscle**	-	Changes the shape of the lens to modify the focal length
-	**Lens**	-	Works with the cornea to focus light on the back of the eye
-	**Cornea**	-	The transparent outside covering on the front of the eye
-	**Retina**	-	A complex layer of nerve cells at the back inside of the eye where the image is projected
-	**Optic nerve**	-	The nerve bundle which transmits visual signals from the retina to the brain
-	**Blind spot**	-	Where the optic nerve exits the retina at the back of the eye
-	**Rods**	-	The dim light receptors located on the retina which have no color sensitivity (More prevalent at the periphery)
-	**Cones**	-	Receptors for direct vision and color discrimination which require good light (More in the center of the retina)
-	**Fovea**	-	The center of the retina where only cone receptors are located (p. 109)
-	**Extrinsic muscle**	-	The six external muscles which control the movement of the eye (p. 110)

7.   How are the following terms used in describing the operations or functions of the human eye?

(HF p. 109-113)

-    **Accommodation**   -   The process by which the refractive power of the eye is modified for appropriate focal length
(HF pp. 109 & 113)

-    **Mandelbaum effect** -   With lack of visual cues the accommodation takes up a empty field or dark focus resting state with a focal length of about 1 meter
(HF p. 113)

-    **Binocular vergence** -   Change in the convergence of the eyes to assist in the focusing or accommodation
(HF p. 113)

-    **Stereopsis**   -   The effect provided by the eye's extrinsic muscle converging the eyes for depth perception
(HF p. 110)

-    **Photopic vision**   -   Color vision from cones in good light

-    **Scotopic vision**   -   Low light rod vision

-    **Adaptation**   -   Adjustment for various light conditions
Coarse adjustment from the pupil More significant process by the rods taking over the low light function

-    **Rhodopsin**   -   The substance called visual purple in the rods which is bleached out in bright light and takes time to be reconstituted

-    **Visual acuity**   -   Refers to the smallest letter which a subject can read on a chart at 20 ft.

- 20/20 is normal vision
- Remains relatively constant up to age 40

- Affected by:
> - Brightness
> - Brightness ratio
> - Contrast
> - Time to view object
> - Glare

(HF p. 112)

PO: 3.   *Describe the elements which make up visual perception.*

8.   What are the main elements of visual perception?

(HF p. 114)

    a.   Eyes

    b.   Inner ear   (Vestibular apparatus) Balancing mechanism

    c.   Brain

    d.   Past experience:

        -        Learning

        -        Expectation

        Refer to Fig 5.6 ----->   (HF p. 115)

9.   What happens when the perception process is denied prediction cues?

(HF p. 114)

    -        Response suffers a delay

10.   Where does the processing of visual information get started?

(HF p. 115)

    -        In the eye itself

11.   What other factors may also play a role in perception?

(HF p. 115)

    -        Emotional factors

12.   Where in the perceptual process do uncertainty and ambiguity occur with correctly sensed information?   (HF p. 115)

    -        In the brain

13. What is **fascination** as it relates to the visual task?

(HF p. 116)

    a. Refers to the failure to respond to a stimulus even though it is within the normal visual field.

    b. Too much concentration on one part of the visual task

    c. A trance-like concentration on a display, receive a message but not respond to it

    d. May be aggravated by fatigue or anxiety

14. How does **set** play a role in visual perception?

(HF p. 116)

    a. When perception is influenced by expectation

    b. One sees what one expects to see

15. How does the **Critical Fusion Frequency** pertain to visual perception?

(HF p. 116)

    - When a flickering light source increases to the point where it is perceived as a steady light

PO: 4.   *Describe what causes the daytime "Blind Spot" and how normal vision compensates for it.*

16. What is meant by the "Blind Spot" in the eye?

(HF p. 115)

    a. Caused by the point the optic nerve exits the retina where there are no rod or cone visual receptors.

    b. Show Fig 5.7 ----->   (HF p. 115)

    c. Binocular vision provides for the other eye to fill in the information missing from the blind spot of the vision of the opposite eye.

    d. Can have serious consequences if the compensating eye's vision is obstructed by a windshield post and the threat object is in the blind spot of the other eye.

PO: 5.   *Describe how the perception of depth and distance are made using each of the various visual cues.*

17.   What are the visual cues the eye and brain uses to establish perception of depth and distance?   (HF p. 117)

      a.   Binocular or stereoscopic vision:

          -    good out to 12 m

      b.   Perspective

      c.   Apparent Movement

      d.   Superposition

      e.   Relative size

      f.   Height of the object in a plane

      g.   Texture gradient

PO: 6.   *Describe the detrimental effects of smoking on vision.*

18.   What are the detrimental affects that smoking has on vision?

                                                (HF p. 117)

      a.   Individuals suffer differently in performance degradation

      b.   Depletion of oxygen supply to the brain

      c.   Altitude and smoking are cumulative

      d.   Serious deterioration in various visual functions:

          1-   Visual acuity

          2-   Brightness threshold

          3-   Reaction to visual stimuli

TERMS OF VISION

LIGHT MEASUREMENT:

- Photometry

- Intensity

- Illumination

- Luminance

- Contrast

- Refraction

- Color

TERMS OF VISION (CONT)

PARTS OF THE EYE:

- Pupil

- Iris

- Ciliary muscle

- Lens

- Cornea

- Retina

- Optic nerve

- Blind spot

- Rods

- Cones

- Fovea

- Extrinsic muscle

TERMS OF VISION (CONT)

FUNCTIONS OF THE EYE:

- Accommodation

- Mandelbaum effect

- Binocular vergence

- Stereopsis

- Photopic vision

- Scotopic vision

- Adaptation

- Rhodopsin

- Visual acuity

ELEMENTS OF VISUAL PERCEPTION

a. Eyes

b. Inner ear

c. Brain

d. Past experience

VISUAL CUES FOR PERCEPTION
OF
DEPTH AND DISTANCE

a.   Binocular or stereoscopic vision:

b.   Perspective

c.   Apparent Movement

d.   Superposition

e.   Relative size

f.   Height of the object in a plane

g.   Texture gradient

HUMAN FACTORS IN AVIATION SAFETY
(SF 320)
LECTURE NOTES

DAY 11   Visual Illusions

PO: 1.   *Be able to identify the ways visual illusions can come into visual perception and give examples of optical and depth/distance illusions.*

1.   Why is it important for pilots and human factors specialists to know about visual illusions?

(HF p. 118)

-    All are vulnerable to being mislead

2.   What are the two areas in which visually sensed information can have errors induced?

(HF p. 118)

a.   Perception process

b.   Interpretation process

3.   What are some of the examples of **optical illusions** common to visual perception?

Refer to Fig 5.8 ----->

(HF p. 118)

4.   What are some examples of **depth and distance illusions** common to visual perception?

Refer to Fig 5.9 ----->

(HF p.119)

PO: 2.   Define the following illusions:

-    Brightness contrast   -   Autokinesis
-    Somatogyral illusion   -   Oculogryal
-    Somatogravic illusion   -   Oculogravic
-    Coriolis illusion   -   Induced movement
                                  -    False horizon

5.   Where is the predominant area of incidents and accidents arising from visual illusions?
                                                                                    (HF p. 122)

        -       Transition to visual control from a non precision (no electronic glide path)
                approach

6.   What causes the **brightness contrast** illusion?

                                                                                    (HF p. 120)

        -       A bright background or surrounding makes the center appear darker

7.   How would you describe the cause of **autokinesis**?   (HF pp. 120 & 123)

        -       An isolated stationary light in an otherwise dark visual field may appear to
                wander

8.   What is the **somatogyral** illusion?

                                                                                    (HF p. 120)

        a.   A false sensation of turning during a prolonged turn

        b.   Caused by the semicircular canals of the vestibular apparatus

9.   What causes the **somatogravic** illusion?

        a.   A false perception of attitude relative to the gravitational vertical

        b.   Caused by the otolith organ of the vestibular apparatus

10.   What are the visual components of the somatogyral and somatogravic illusion called?
                                                                                    (HF p. 120)

        -       **Oculogyral** and **oculogravic**

11.   How does the cross coupled or **Coriolis** illusion occur?   (HF p. 120)

        a.   Motion in the fluid of two semicircular canals stimulates motion in the third

        b.   Can be caused by turning the head while already in a turn

        c.   Gives a sense of tumbling

12.   What causes the illusion of **induced movement**?   (HF p. 120)

        -       Visual stimuli from a moving rather than a stationary subject

13.   What can cause the perception of **false horizon**?   (HF p. 125)

        a.   Surface lights confused with stars

        b.   Sloping cloud layers or Aurora Borealis

PO: 3.   Describe the various illusions which can occur during pilot operations of aircraft.

14.   What are some examples of illusions that can occur during taxying? (HF pp. 124-125)

        a.   Loading bridge moving away   -   Induced movement

        b.   Blowing snow   -   Induced movement

        c.   Pilot eye height   -   Reduced apparent motion

15.   What are the illusions characteristic to the takeoff phase of operations?   (HF p. 125)

        a.   Sensation of pitching up during acceleration

                -       Somatogravic and oculogravic

        b.   Perception of false horizon

        c.   Sloping terrain   -   excessive or insufficient height

16.   To what illusions might a pilot be susceptible in cruising flight?

                                                                              (HF p. 126)

        a.   Holding pattern = prolonged turning   -   somatogyral

                -       Coriolis if head is turned

        b.   An isolated light against a dark background -   Autokinesis

c.   Illusions of relative altitude:

- Approaching aircraft appear to be higher at first

- Mountains appear at first above the horizon

17.   What are some of the conditions which contribute to the illusions pilots can experience during the approach and landing phases of operations?

(HF p. 126)

a.   Fatigued at the end of a long flight

b.   Pressure of the terminal area

c.   Bad weather or poor visibility

18.   What are some of the approach and landing illusions which can affect pilots?

Refer to Fig 5.15 ----->   (HF pp. 127-128)

a.   Sloping terrain to the runway:

Condition	Illusion	Effect
- Up slope	= too high	too low
- Down slope	= too low	too high

b.   Sloping runway:

Condition	Illusion	Effect
- Up slope	= too steep	too shallow
- Down slope	= too shallow	too steep

c.   Runway width different than normal:

Condition	Illusion	Effect
- Wider	= too low	too high - dropped in
- Narrower	= too high	too low - late round out

d. Light intensity:

Condition   Illusion

- Brighter   = closer

- Dimmer   = further

e. Visibility restriction (Mist or fog):

(HF p. 123)

Condition   Illusion

- Brighter   = closer

- Dimmer   = further

f. Windshield Location - pitch angle:

Condition	Illusion
- Low pitch - high speed	object is higher
- High pitch - low speed	object is lower

g. Runway texture:

- Lack of texture - reduces depth perception

- Glassy water surface

- Snow covered runway

- Dimly lighted surfaces at night

PO: 4. Describe five things which can be done to minimize flightcrew susceptibility to visual illusions.

19. What are five things flightcrews can do to reduce their chance of experiencing visual illusions? (HF p. 129)

a. Recognize them as a normal phenomena

b. Understand the nature of the situations in which they are likely to occur

c.   Supplement visual cues with information from other sources

-    Protective use of back up aids:

-    radio aids
-    radar
-    attitude displays
-    radio altimeter
-    distance measuring equipment (DME)
-    inertial and other navigational systems
-    visual glide slope aids (VASI)

-    Non precision approaches have more accidents from visual illusions

-    Take special care when the crew is tired or when visual cues are reduced

d.   Note on airport and approach charts geographic locations which are known to be associated with visual illusions

e.   Manufacturers and certifying authorities must insure:

-    Adequate flight deck visibility for both vision angle and optical quality of the glass

PO: 5.   Define the "design eye" in terms of cockpit geometry

20.   From where does the standard for the required visibility envelope for the flight deck of civil aircraft come?   (HF pp. 120-121)

-    Society of Automotive Engineers (SAE) (SAE AS 590)

21.   Where should the "design eye" be located for civil transport aircraft?   (HF p. 121)

a.   To allow the pilot to see a length of approach or touch-down zone lights which would be covered in three seconds at final approach speed

b.   Usually about 200-250 m (548-685 ft) along the flight path

22.   What sitting height position will place the pilot's eye at approximately the "design eye" location?   (HF p. 122)

-    The upper surface of the glare shield should provide a lateral horizontal reference

PO: 6.   Describe some problems which result from poor cockpit window design.

23.   How can the optical characteristics of windows affect visibility?

(HF p. 124)

a.   Distortion from poorly designed or shaped glass

b.   Delamination bubbles can form with wear before replacement

c.   Cockpit blind spots can result from windshield posts

VISUAL ILLUSIONS

- Brightness contrast

- Autokinesis

- Somatogyral illusion   -   Oculogryal

- Somatogravic illusion   -   Oculogravic

- Coriolis illusion

- Induced movement

- False horizon

Sloping terrain to the runway:

Condition	Illusion	Effect
- Up slope	= too high	too low
- Down slope	= too low	too high

Sloping runway:

Condition	Illusion	Effect
- Up slope	= too steep	too shallow
- Down slope	= too shallow	too steep

Runway width different than normal:

Condition	Illusion	Effect
- Wider	= too low	too high - dropped in
- Narrower	= too high	too low - late round out

TO REDUCE THE CHANCE
OF
VISUAL ILLUSIONS

1. Recognize them as a normal phenomena

2. Understand likely situations

3. Supplement visual cues

4. Note locations which are known to be associated with visual illusions on charts

5. Insure adequate flight deck visibility for both vision angle and optical quality of the glass

HUMAN FACTORS IN AVIATION SAFETY
(SF 320)
LECTURE NOTES

DAY 12   Motivation and Safety

PO: 1.   *Identify when human behavior began to be an important part of American accident investigation.*

1.   What is the reason that a properly qualified, highly trained, medically fit, well paid person would fail to perform the task expected?

(HF p. 131)

    a.   It takes more than finding out **WHAT** happened

    b.   We must find out **WHY** it happened

    c.   Perhaps there must be an examination of psychological factors such as motivation.

2.   When did the National Transportation Safety Board (NTSB) first establish a separate Human Performance Division?   (HF p. 131)

    -    in 1983

3.   What is necessary in order to accomplish the best matching of the Liveware component to all the other components of the SHEL model?

(HF p. 132)

    -    Recognition of human:

        1-   Characteristics

        2-   Capabilities and limitations

        3-   Behavior patterns

PO: 2.   *Define what is meant by the term "motivation" and describe the different levels of motivation.*

4.   What is probably the most significant characteristic of the Liveware component in driving a person to behave in a particular way?   (HF p. 132)

-      Motivation

5.   How would you define motivation?   (HF p. 132)

a.   The internal force which initiates, directs, sustains and terminates all important activities.

b.   The difference between what a person **CAN** do and what a person **WILL** do

c.   People are different and are driven by different motivational forces

d.   Influences the level of performance, the efficiency achieved and the time spend on an activity

6.   Where does the basic level of motivation begin?   (HF p. 133)

a.   As a need or requirement for survival

-      Individual:

-      Air   -   Water   -   Food

-      Basic shelter

-      Species:

-      Sexual drive

b.   The relative strength of these drives determines the direction of our motivation and activities.

7.   What is the end result of a sequence of motivated behavior called?

     (HF p. 132)

-      Goal or purpose

8.   What determines the way we go about satisfying needs?   (HF p. 133)

      -      Depends on previous experience

             -      Child -   reward and punishment   -   Externalize

             -      Adult -   more complex and subtle   -   Internalized

9.   What may be the result of setting an unrealistically high goal?

                                                                      (HF p. 133)

      a.   Change to a secondary goal and adjust the motivation there

      b.   May become frustrated, irritable and neurotic

10.   How important is organizational culture to the motivation required for crews to
      maintain high standards of professionalism and exercise proper discipline?
                                                                      (HF pp. 134-136)

      -      The NTSB recognizes it as causal in the chain of circumstances leading up to
             an accident.

11.   What some aspects of an individual's motivational system?

                                                                      (HF pp. 136-137)

      a.   Sets into motion a sequence of behavior

      b.   Drive leads to goal directed activity

      c.   Most fundamental drives are physiological

      d.   Then psychological or social orientation

PO: 3.   *Describe the three basic theories of motivation, and show how Maslow's theory*
                on the hierarchy of needs works in relation to human motivation.

12.   What are the three basic theories of motivation?

                                                                      (HF p. 137)

      a.   Learning Theory:

             -      Relates present motivation to past experience

             -      A rewarded response will be repeated

126

b.   Psychoanalytic Theory:

-      Modified to use ego or self concept

c.   Cognitive Theories:

-      Emphasize the human as a rational being free to make choices as a result of perception, thought and judgement

-      Goal directed behavior is a result of:

  -      Attitudes, beliefs, values and expectations

13.   How does Maslow's theory on the hierarchy of needs work in relation to human motivation?

Refer to Fig 6.1 ----->   (HF p. 138)

a.   It starts with basic physiological needs

b.   Goes through the psychological and social needs

c.   The final needs to meet are the self fulfillment needs (self actualization)

PO: 4.   *Describe the contributions of the industrial studies of Taylor, Hawthorne, and the Two-Factor Theory of Herzberg to understanding the motivation behind behavior.*

14.   What do we learn about motivation from the industrial studies of Taylor, Hawthorne, and the Two-Factor Theory of Herzberg?   (HF p. 138-139)

a.   Taylor -   1890s at Bethlehem Steel Company

-      Money was considered the prime motivating factor

b.   Hawthorne studies - 1920s at Western Electric

-      Modifying the factory lighting improved work performance

-      Work was influenced by social and psychological factors quite independent of the work itself

-      Management's care about working conditions

c.   Two-Factor Theory of Herzberg - 1959

- Satisfaction came from motivating factors:

- achievement
- advancement
- recognition for good work
- responsibility
- the nature of work itself

- Dissatisfiers or hygiene factors:

- staff relations
- company personnel policy
- salary
- security
- poor working conditions

PO: 5.   Describe how Murray's Motives fit into the theories of motivation.

15.   What are the premises which form the foundation of Murray's Motives?   (HF p. 139)

a.   There are separate, distinguishable drives which control behavior

b.   Some form of hierarchy may determine the order of priority which is put on them

c.   People may be motivated by different forces while apparently behaving in the same way

16.   How would you describe the three extensively investigated motives cited by Murray?
(HF pp. 139-140)

a.   **Achievement motivation:**

- Made up of three distinct elements:

1-   Mastery -   need to confront new challenges and surpass earlier performance

2-   Work   -   reflecting the satisfaction gained from performance

3-   Competitiveness   -   drive to surpass the performance of others

- Drive towards achievement for its own sake rather than material benefits

- Strive to meet or surpass set standards

- Sets high standards

- Works efficiently without supervision

- May be bored with routine tasks

- Looks for good feedback on the job

- These traits are highly stable and modification is not easy

- Frequently associated with leadership

- Paramount importance in jobs with:

    - a high level of unsupervised performance
    - involving initiative
    - responsible, well-disciplined behavior

b. **Affiliation motivation:**

- Concerned with the establishment and maintenance of affectionate relationships   (HF p. 140)

- Desire to be liked and accepted by people

- Involves adherence and loyalty to a friend

c. **Power motivation:**

- Concern over the means of influencing the behavior of another person

17. What happens if it is impossible to satisfy these individual motivational needs?
                                                                        (HF p. 141)

a. Aggressive behavior

b. Fighting the "system"

c. Withdrawing cooperation

      d.   Developing intense dislikes for individuals or groups who are seen as being responsible

PO: 6.   *Describe some of the considerations of the concept of expectancy and rewards in motivation.*

18.   In what way did Vroom criticize the Two-Factor Theory of Herzberg?   (HF p. 141)

      a.   He thought the theory was too simple and did not take into account individual differences

      b.   He included individual preferences for pay, promotion or security

      c.   He took into account expectancy - or the chance a person sees that his behavior or performance will lead to the desired outcome.

      d.   The Vroom model was not a very practical tool for management use as it did not provide guidelines on what motivates the staff at work

19.   How does expectancy of rewards fit into motivation?

          Refer to Fig 6.2 ----->   (HF p. 141)

      a.   A rewards usefulness is seen as the product of:

                                                   (HF p. 142)

          -     The value placed upon it

          -     The probability of attainment

      b.   Effort is not synonymous with performance

      c.   Two other variables must be applied to achieve:

          -     Natural abilities

          -     Learned skills

20.   What is the most important aspect of the availability of rewards?   (HF p. 142)

      a.   What the employee **sees** as available

      b.   Not what management **says** is available

c. The high performer must see that high performance is rewarded more than low performance

PO: 7. *Describe factors which influence job satisfaction, and differentiate between job enlargement and job enrichment.*

21. What are some of the factors which influence the **job satisfaction** part of motivation?
(HF p. 142)

   a. Financial rewards

   b. Management personnel policies

   c. Work colleagues

   d. The working environment

   e. The nature of the task itself

22. How does job satisfaction tie into performance? (HF p. 143)

   a. It is not necessary for, nor does it automatically result in improved performance

   b. Little evidence from industry showing increasing job satisfaction results in higher performance unless rewards are tied to performance and are seen to be tied to it

   c. More dissatisfaction with relative income lower than another of comparable job than absolute income too low

   d. Incentives can become institutionalized as rights rather than rewards for good performance

   e. A rationally critical attitude of an employee towards the supervisory or management policies should not be confused with job dissatisfaction or low motivation

23. What is the difference between the job enlargement and job enrichment approaches to increasing job satisfaction? (HF p. 144)

   a. Job enrichment:

      - Active participation of staff in policy and decision-making concerning their work

131

b. Job enlargement:

- Increasing the number and variety of tasks (horizontal enlargement)

- Increasing the person's control of the routine planning of the task (vertical enlargement)

c. Both have shown promise in industry but are not the complete answer in and of themselves

24. What guidance has been given for having a well established relationship between goal achieving and job satisfaction? (HF p. 144)

a. Goals and Targets should be:

- Clear, precise and acceptable

- Realistic but have some degree of challenge attached

- Given with established acceptable tolerances

b. These are very effective in enhancing performance for most tasks which are related to safety and efficiency

25. What is one of the most pervasive problems facing advanced industrial societies? (HF p. 144)

- Boredom

26. What are the six categories, suggested by surveys of job attitudes, into which boredom may fall? (HF p. 144)

a. Constraint

b. Meaninglessness

c. Lack of interest

d. Repetitiveness

e. Lack of sense of completion

f. The perception of the jobs general nature

f.   The perception of the jobs general nature

PO: 8.   *Describe the difference between positive and negative methods of reinforcing or discouraging behavior.*

27.   What is the difference between positive and negative behavior reinforcement?

(HF p. 144)

a.   Positive reinforcement:

-     Strengthening the desired behavior by rewards

b.   Negative reinforcement:

-     Discouraging the undesired behavior by punishments

c.   Positive reinforcement is more effective in raising performance

d.   Reinforcement should closely follow the intended behavior

28.   What are some of the precautions that must be applied when one is required to use negative reinforcement?

(HF p. 145)

a.   Don't use for errors caused by lack of skill or knowledge

b.   Group punishments may be counter-productive

c.   Petty penalties for minor indiscretions seem ineffective

d.   Must be administered with complete fairness to all offenders

PO: 9.   *Describe the elements of crew resource management (CRM).*

29.   How important is it for good leadership in the management of crew resources?

(HF p. 146)

-     Research of accident and incident data present a powerful case for improved understanding and application of crew resource management (CRM)

30. What is involved in CRM? (HF p. 146)

    a. Leadership

    b. Group relations

    c. Communications

    d. Motivation

PO: 10. Define the role of leadership, and describe the characteristics and tasks of a leader.

31. What is a leader? (HF p. 146)

    a. A person whose ideas and actions influence the thought and behavior of others

    b. An agent of change and influence

32. What are the tasks of the effective leader? (HF pp. 146 & 148-149)

    a. Understand the group goals

    b. Understand the company goals

    c. Feel responsible for implementing these goals

    d. Motivating the members of the group

        - Emphasizing the objectives

        - Clarifying the targets or goals which should be achieved

    e. Modifying habits and behavior by reinforcement

    f. Demonstration of the desired goals and behavior by example

    g. Maintaining the group:

        - Attending to personal relationships

        - Resolving differences

        - Encouraging harmony and cooperation

h.   Managing the total resources:

-        Allocation of duties

-        High stress situations - group is more dependent

33.   What is the difference between leadership and authority?   (HF pp. 146-147)

a.   Authority is normally assigned

b.   Leadership is acquired and suggests voluntary following

c.   Optimally -   authority is combined with true leadership

34.   In what ways do all members of a group contribute to the effective leadership of the
group?   (HF p. 147)

a.   Supplying information

b.   Contributing ideas

c.   Providing support

d.   General response to the leader

35.   What are the characteristics of a leader?   (HF p. 147)

a.   Technical skills

b.   Judgement and intelligence

c.   Demonstrated achievement

d.   Responsibility

e.   Ability to participate and cooperate with others

f.   Understand group needs and objectives

MOTIVATION & LEADERSHIP

MOTIVATION

- The internal force which initiates, directs, sustains and terminates all important activities.

- The difference between what a person <u>CAN</u> do and what a person <u>WILL</u> do

- Probably the most significant characteristic of the LIVEWARE component

- Influences the level of performance, the efficiency achieved and the time spend on an activity

BASIC LEVEL MOTIVATION

A need or requirement for survival

- Individual:

    - Air   - Water   - Food

    - Basic shelter

- Species:

    - Sexual drive

Motivated Behavior -----> Goal or Purpose

- Depends on previous experience

    - Child - reward and punishment - Extrinsic

    - Adult - more complex and subtle - Intrinsic

MOTIVATIONAL SYSTEM

- Sets into motion a sequence of behavior

- Drive leads to goal directed activity

- Most fundamental drives are physiological

- Then psychological or social orientation

THEORIES OF MOTIVATION

Learning Theory:

- Relates present motivation to past experience

- A rewarded response will be repeated

Psychoanalytic Theory:

- Modified to use ego or self concept

Cognitive Theories:

- Emphasize the human as a rational being free to make choices as a result of perception, thought and judgement

- Goal directed behavior is a result of:

  - Attitudes, beliefs, values and expectations

RESEARCH STUDIES

Taylor - 1890s at Bethlehem Steel Company

- Money was considered the prime motivating factor

Hawthorne studies - 1920s at Western Electric

- Work was influenced by social and psychological factors

- Management's care about working conditions

Two-Factor Theory of Herzberg - 1959

- Satisfaction came from motivating factors:

  - achievement
  - advancement
  - recognition for good work
  - responsibility
  - the nature of work itself

- Dissatisfiers or hygiene factors:

  - staff relations
  - company personnel policy
  - salary
  - security
  - poor working conditions

THE PREMISES FOR

MURRAY'S MOTIVES

a.   There are separate, distinguishable drives which control behavior

b.   Some form of hierarchy may determine the order of priority which is put on them

c.   People may be motivated by different forces while apparently behaving in the same way

MURRAY'S MOTIVES

a.   Achievement motivation

b.   Affiliation motivation

c.   Power motivation

Achievement motivation:

- Made up of three distinct elements:

    1- Mastery

    2- Work

    3- Competitiveness

- Paramount importance in jobs with:

    - a high level of unsupervised performance

    - involving initiative

    - responsible, well-disciplined behavior

JOB SATISFACTION COMES FROM

a. Financial rewards

b. Management personnel policies

c. Work colleagues

d. The working environment

e. The nature of the task itself

FOR MORE JOB SATISFACTION

- Goals and Targets should be:

    - Clear, precise and acceptable

    - Realistic but challenging

    - Given with established acceptable tolerances

CAUSES OF BOREDOM

a. Constraint

b. Meaninglessness

c. Lack of interest

d. Repetitiveness

e. Lack of sense of completion

f. The perception of the jobs general nature

REINFORCEMENT

Strengthening   Discouraging

POSITIVE   NEGATIVE

<u>Rewards</u>   <u>Punishments</u>

ELEMENTS
OF
CREW RESOURCE MANAGEMENT
(CRM)

a. Leadership

b. Group relations

c. Communications

d. Motivation

EFFECTIVE LEADER TASKS

a. Understand the group goals

b. Understand the company goals

c. Feel responsible for implementing these goals

d. Motivating the members of the group

-   Emphasizing the objectives

-   Clarifying the goals

e. Modifying habits and behavior by reinforcement

f. Demonstration of the desired behavior by example

g. Maintaining the group:

-   Attending to personal relationships

-   Resolving differences

-   Encouraging harmony and cooperation

h. Managing the total resources:

-   Allocation of duties

-   High stress situations - group is more dependent

LEADERSHIP

Leader:

a.   A person whose ideas and actions influence the thought and behavior of others

b.   An agent of change and influence

AUTHORITY VS LEADERSHIP

a.   Authority -----> Assigned

b.   Leadership ----> Acquired

CHARACTERISTICS
OF
A LEADER

a. Technical skills

b. Judgement and intelligence

c. Demonstrated achievement

d. Responsibility

e. Ability to participate and cooperate with others

f. Understand group needs and objectives

DAY 13   Communication

PO: 1.   *Give the definition and some examples of types of communication.*

1.   How important is effective communication to our society and to the aviation industry?
(HF p. 152)

    a.   Social, economic and technological efficiency all depend on effective communication

    b.   Loneliness, distress and death in the aged without it

    c.   Industrial strikes, and aircraft crashes result without it

2.   What is the definition of communication?

    -    giving or exchanging of information

3.   Where does communication fit in the SHEL conceptual model?   (HF p. 152)

    -    At any and all interfaces between the components

4.   What are some of the types of communications which must be optimized for efficiency and safety?   (HF p. 152)

    a.   Verbal:

        -    Spoken

        -    Written

        -    Keyboard

    b.   Non-verbal:

        -    Body language

        -    Symbols, characters and diagrams

c.   Para-verbal

        -       Grunts groans and moans

d.   Electronics

        -       machine ---> human

        (voice synthesizers)

        -       human ---> machine

        (keyboard)

        (automatic speech recognition) (ASR)

PO: 2.   *Describe the elements required for communication and the place language has in the process.*

5.   What is required to accomplish communication?

Sender ---> Message ---> Medium ---> Receiver

I_____   Language _____I

6.   Which of the elements of the communication process uses language?   (HF p. 153)

    a.   It is part of the message and medium between the sender and the receiver

    b.   It is also a part of the cognitive process of both the sender and the receiver

7.   What is ambiguity as it relates to the language of communication?   (HF p. 153)

    a.   A characteristic of communication which varies with the different form of language

    b.   Sometimes speaking clarifies meaning of words with the same appearance:

        -       Tear
        -       Lead
        -       "They are eating apples"

c. Sometimes writing clarifies meaning of words sounding the same:

- Red vs Read
- Lye vs Lie

d. Sometimes only context can clarify the meaning of a word with the same sound and appearance:

- Don't tell a **lie**   vs   **Lie** down and take it
- A warning **light**   vs   **Light** things don't weigh much

8. What are the principles or rules called which govern the arrangement of words in language? (HF p. 154)

- Syntax or grammar

9. What are the differences in the roles of written and spoken language? (HF p. 154)

a. Speech
- Provides rapid exchange of messages
- Primary lubricant of social interaction
- Uses the auditory channel
- Uses pronunciation and accent for clarity

b. Writing
- Characteristic of permanence in time and space
- Information accumulation and storage
- Uses the visual channel
- Uses punctuation and accent for clarity

10. How does body language fit into the communication picture? (HF pp. 154-155)

a. There is a vast range of body language

b. It has a surprising degree of universality in its interpretation

c. Is much influenced by culture

d. Spatial proximity also influenced by culture and personality

- Introverts need more space than extroverts

PO: 3.   *Define the term intelligibility in spoken language and give examples of ways to increase it.*

11.   What is meant by the term intelligibility in spoken language?   (HF p. 155)

-     The extent to which the transmitted word is understood by the listener

12.   What are some of the factors that make a word more intelligible?   (HF p. 155)

a.   Familiarity:

-     The frequency with which the word is used in every day life the easier it is understood

b.   Length:

-     Generally, the longer the word, the more readily it is identified

-     Sometimes length "Negative" is easier to catch than the familiar "No"

c.   Context:

-     A word in a phrase or a sentence is more likely to be understood than a word by itself

d.   Repetition:

-     Usually increases the understanding of a word

13.   What has been done to use these principles to improve communications in the aviation industry?   (HF pp. 155-156)

a.   Adoption of standard phraseology in aircraft communications

b.   Standard word spelling and phonetic alphabets

-     Increase intelligibility when communication conditions are poor

-     Reduce the risk of misunderstanding at all times

c.   Although much of the research which led to the standard international vocabulary and phraseology originated in the U.S. we seemed to be behind now.

14. How does the standard vocabulary effect the intelligibility of a message?

        Refer to Fig 7.1 ----->   (HF p. 156)

    a.   Small standardized vocabulary is best

    b.   Familiar words are next best

    c.   Large vocabulary of different words is the least

PO: 4.   Describe the main parts of the vocal and auditory systems and their functions.

15. What are the main parts of the vocal system?

        Refer to Fig 7.2 ----->   (HF p. 157)

    THE SENDER --->   Vocal system

    -        Originates the sounds (Vibrations) of vocal communication

    a.   Lungs

    b.   Trachea

    c.   Larynx

    d.   Pharynx

    e.   Nose

    f.   Mouth

16. What are the main parts of the auditory system?

        Refer to Fig 7.3 ----->   (HF p. 158)

    THE RECEIVER --->   Auditory system

    -        Senses and conveys vocal communication to the brain

a.   Outer ear

- External auditory canal

- Ends at the tympanic membrane (eardrum)

b.   Middle ear

- Ossicles: hammer, anvil and stirrup

- Sets in motion the fluid of the inner ear

c.   Inner ear

- Cochlea and the Organ of Corti

- Contains the nerve ends and hair cells

- Act as transformers and amplifiers

- Semicircular canals

- Organs of balance and orientation

d.   Brain

- Receives and interprets the signals from the inner ear

PO: 5.   Describe the types of conditions which can lead to loss of hearing ability.

17.   What are the three ways hearing loss can occur?   (HF pp. 158-159)

a.   Conduction deafness

b.   Nerve deafness

c.   Central hearing loss

18.   What is the interference with the transmission of sound waves through the outer and middle ear are called?   (HF p. 159)

- Conduction deafness

19. What are some of the things which can cause hearing deficiencies from conduction deafness? (HF pp. 158-159)

  a. Outer ear :

    - Wax accumulation

    - Pressure differential from blocked eustachian tube

  b. Middle ear:

    - Deposits of new bone (calcium) on the ossicles (Otosclerosis)

    - Infections of the middle ear (more common) (Otitis media)

20. What are the conditions which can cause nerve deafness? (HF p. 159)

  a. Damage to the hair cells and nerve fibers of the inner ear

  b. Noise is the biggest threat

    - From a single loud noise

    - From long-term exposure to noise

  c. In both cases the cells suffer permanent damage

  d. Deafness usually advances gradually and insidiously

21. What causes a central hearing loss? (HF p. 159)

    - Interference with the functioning of the brain which is associated with hearing

22. What happens to hearing as a result of age?

    Refer to Fig 7.4 -----> (HF p. 160)

  a. Deteriorates with age especially in the higher frequencies

  b. More severe in men than women

PO: 6.　　*Describe the four characteristics of speech which influence intelligibility.*

23.　What are the characteristics of speech which can have an affect on the intelligibility of the message?　　　　　　　　　　　　　　　(HF pp. 160-161)

　　　a.　Intensity　　　-　　　Measured in decibels　　　(loudness)

　　　　　　-　High frequency sounds louder than low frequency

　　　　　　-　Vowels sound louder than consonants

　　　　　　-　Consonants carry a large part of the information in speech with most languages

　　　b.　Frequency　　　-　　　Measured in Hertz　　(pitch)

　　　　　　-　Healthy human ear from 16 Hz - 20,000 Hz

　　　c.　Harmonic composition　　　　　　　　(quality)

　　　　　　-　Can change meaning

　　　　　　(from sympathetic to sarcastic)

　　　d.　Speed

　　　　　　-　Length of pauses

　　　　　　-　Time spent on the different sounds

PO: 7.　　*Describe how clipping, masking, and noise affect the spoken message.*

24.　What are the kinds of degradation which can effect the reception of the spoken message?
　　　　　　　　　　　　　　　　　　　　(HF p. 161)

　　　a.　Clipping:

　　　　　　-　Several frequency bands are cut out

　　　　　　-　Intermittent total cuts in speech

　　　　　　-　Because of the quality of redundancy in speech the message usually still gets through

　　　　　　-　Increases the chance for expectation and other errors

b.   Masking:

- Unwanted noise from the environment

- Radiomagnetic interference

- Most effectively protection is to isolate or control it at the source

- Consonants carry most information but are most vulnerable to masking

- Signal-noise ratio = relationship between the signal loudness and the background noise

- Increasing the volume of both the signal and the background noise will do little to enhance the message

c.   Noise:

- Sound which has no relationship to the completion of the immediate task

- Control at the source and at the receiver

25.   How does hearing protection reduce the affects of noise?   (HF p. 162)

a.   Muffs and plugs protect from damage of the hair cells of the inner ear

b. Does not degrade speech intelligibility because the
signal-noise ratio remains the same

26. How do visual cues combined with auditory information affect message reception?

Refer to Fig 7.5 ----->   (HF p. 162)

a.   Visual cues always assist the message

b.   The higher the speech-to-noise ratio, the greater the advantage of visual cues

(next page)

PO: 8.   *Explain how expectation can influence the reception of the spoken message and what can be done to reduce this phenomenon.*

27.   What increases the risk of expectation causing an error in understanding the spoken message?   (HF p. 163)

-       The amount of the speech content lost due to:

        -       Clipping

        -       Distortion

        -       Noise

        -       Personal hearing loss

28.   Where is the phenomenon of expectation particularly common and dangerous, and what can be done about it?   (HF p. 163)

a.   In the read-back and confirmation of messages

b.   Short term memory deficiencies seem to compound the problem

c.   A read-back confirmation can reduce the error if it is not done in a perfunctory fashion

d.   Avoid giving more than one altitude in a clearance will help

f.   Data link with visual display is helpful to reduce errors

g.   Wording positive and negative messages differently

h.   If in doubt request a repeat

i.   Involve more than one crew member on important messages

j.   Place verbal stress on critical words

PO: 9.   *Describe some of the means of measuring the effectiveness of spoken language.*

29.   What is the use of an articulation index good for?   (HF p. 164)

a.   Used to compare the efficiency of different communication systems

b.   Converted into intelligibility =

- The percentage of spoken material of any particular type which is understood by a listener

30.   What is meant by the Speech Interference Level?   (HF p. 165)

a.   The destructiveness of noise on the reception of speech

b.   Used to predict the voice level required for communication at certain noise levels

PO: 10.   Describe some of the considerations in the development and use of Automatic Speech Recognition.

31.   What is Automatic Speech Recognition (ASR), and what are some of the problems to be overcome in its development?   (HF p. 165)

a.   Computer voice recognition

b.   Difficulty in breaking human speech down into sequences of discrete words

c.   Different people speak words differently

- accents

d.   Harmonic composition or quality of speech can vary

- Illness, Fatigue, or emotion

e.   Requires many levels of complex processing

f.   Most are speaker-dependent

32.   What are the three main issues which must be resolved in order to have effective ASR systems?   (HF p. 166)

a.   Must respond to input from a wide variety of users

b.   Ways to ignore background noise must be found

c.   Words which are stored separately must be connected together in a meaningful way

33. What is the essential design criterion for ASR to be used on the flightdeck?

(HF p. 166)

- Reliability

PO: 11. *Explain some of the safety issues associated with radio communications, and describe some changes for its improvement.*

34. Why is it important for all pilots to use and understand a common language?

(HF p. 166)

a. To be able to follow instructions or communicate needs

b. To monitor the instructions and requests for other aircraft using the same airspace

35. What impact do errors in understanding oral communication play in aviation safety?

(HF p. 167)

a. About 70% of the ASRS reports involve some kind of oral communication problem related to the operation of aircraft

b. Expectation was very prominent as a determined cause

36. What was the most critical error centered around in the Tenerife double 747 accident?

(HF p. 168)

- The meaning of the word "cleared"

    - KLM took it for both takeoff and route clearance

37. What changes in radio phraseology were made after the Tenerife double 747 accident?

(HF p. 170)

a. Restricted use of the words "clear" and "clearance" and "takeoff"

    - "Departure" replaced for "takeoff" except for takeoff itself

    - "Clear" only used for takeoff, route, approach and landing clearances

b. Changed "affirmative" to "affirm"

c. Deleted "this is", "over", and "out"

COMMUNICATION

- giving or exchanging of information

Requirements are:

Sender ---> Message ---> Medium ---> Receiver

I_____ Language _____I

- inextricably linked ---> the cognitive process

and

---> communication

TYPES OF COMMUNICATION

- Verbal

- Non-verbal

  (body language)

  (voice inflection)

- Written

- Symbols and diagrams

- Electronics

  - machine ---> human

    (voice synthesizers)

  - human ---> machine

    (keyboard)

    (automatic speech recognition) (ASR)

163

LANGUAGE

SPEECH   ---> Auditory channel

WRITTEN   ---> Visual channel

BODY LANGUAGE

-      surprisingly universal

-      much influenced by culture

Intelligibility

-      the extent to which the transmitted word is understood by the listener

ENHANCING INTELLIGIBILITY

- Familiarity:

- Length:

- Context:

- Repetition:

- Standardized phraseology and phonetic alphabet

SPOKEN MESSAGE

THE SENDER ---> Vocal system

- Lungs

- Trachea

- Larynx

- Pharynx

- Nose

- Mouth

THE RECEIVER ---> Auditory system

- Outer ear

- Middle ear

  - Ossicles: hammer, anvil and stirrup

- Inner ear

  - Cochlea and the Organ of Corti

  - Semicircular canals

- Brain

HEARING LOSS

a. Conduction deafness

b. Nerve deafness

c. Central hearing loss

SPEECH CHARACTERISTICS

1- Intensity - Measured in decibels (loudness)

- High frequency sounds louder low frequency

- Vowels sound louder than consonants

2- Frequency - Measured in Hertz (pitch)

- Healthy human ear from 16 Hz - 20,000 Hz

3- Harmonic composition (quality)

- Can change meaning

(from sympathetic to sarcastic)

4- Speed

- Length of pauses

- Time spent on the different sounds

BARRIERS TO VOCAL COMMUNICATION

- Clipping:

    - Several frequency bands are cut out

    - Intermittent total cuts in speech

- Masking:

    - Unwanted noise from the environment

    - Consonants carry most information

        - Also most vulnerable to masking

- Noise:

    - Sound which has no relationship to the completion of the immediate task

MAIN ISSUES
OF
AUTOMATIC SPEECH RECOGNITION (ASR) SYSTEMS

a. Must respond to input from a wide variety of users

b. Ways to ignore background noise must be found

c. Words which are stored separately must be connected together in a meaningful way

HUMAN FACTORS IN AVIATION SAFETY
(SF 320)
LECTURE NOTES

DAY 15   Attitudes & Persuasion

PO: 1.   *Define and describe the differences between personality traits, attitudes, beliefs, and opinions.*

1.   How would you define a personality trait?   (HF p. 172)

  a.   Deep-seated characteristics which constitute the essence of a person

     -    Sometimes have a tendency to alter in middle age

  b.   Are stable and very resistant to change

     -    Attempts to modify personality distortions have only limited success

  c.   Are far more complex than can be encapsulated in any conventional personality test

  d.   Some traits distinguish pilots from the rest of the population

     -    Anxiety state is the most common psychiatric disorder among those flying professionally

     -    Different kinds of flying require different types of personality

     -    More research is necessary into desirable and undesirable personality characteristics related to modern civilian flying and flight deck management.

  e.   Need to assess during pilot selection and initial training

2.   How would you define an attitude?

    (HF p. 173)

  a.   Likes and dislikes

  b.   A learned and rather enduring tendency to respond favorably or unfavorably to people, decisions, organizations or other objects

  c.   A predisposition to respond in a certain way

  d.   Not a response or behavior other than expression of an opinion

171

3.   What is a belief?   (HF p. 173)

    a.   Does not necessarily infer a favorable or unfavorable evaluation

    b.   An assertion about two objects or the relationship between them

4.   How would you describe an opinion?   (HF p. 173)

    -      A verbal expression of an attitude or belief

5.   How can personality, attitudes and beliefs be studied?   (HF p. 173)

    a.   They are intangible and can't be studied directly

    b.   We can only infer them from observed behavior

6.   In the jargon of psychology, what would the terms personality, attitudes and beliefs be called?   (HF p. 173)

    -      Hypothetical constructs

PO: 2.   *Define the developmental influences for, and the components of attitude.*

7.   Where do attitudes have their origins?   (HF p. 174)

    a.   In early life experiences

        -      Family

        -      Political orientation

    b.   In the social environment

    c.   The media

8.   What are the three components of attitude?   (HF p. 174)

    a.   Cognitive   -   knowledge, idea or belief about a subject

    b.   Affective   -   feelings held about it

    c.   Behavioral   -   what is said or done about it

172

9.   What might be some of the difficulties in evaluating attitudes by behavior?

(HF p. 174)

    a.   May be inconsistency between verbal expression of the attitude and other behavioral expressions

    b.   The apparent paradox arises because behavior is determined by more than one attitude

PO: 3.   *Describe how stereotyping can effect the development of attitudes.*

10.   What is stereotyping in terms of attitudes?   (HF p. 175)

    -      The tendency to categorize or classify things or people

11.   Why do people tend to use stereotypes?   (HF p. 175)

    -      We don't have the time or the inclination to make a strictly individual analysis before adopting attitudes about them

12.   What is the danger of stereotyping?   (HF p. 176)

    a.   It is unjustified even when based on personal experience

    b.   It is frequently based on rumor or word of mouth

    c.   Often it is based on unsound generalizations or prejudices

    d.   It sets us up for the error of expectancy and we tend to see and hear what we expect

PO: 4.   *Describe some of the reasons for having attitudes.*

13.   What are some of the more common reasons for having attitudes?   (HF p. 177)

    a.   Provide some sort of cognitive organization of the world

    b.   Give a structure which facilitates thinking and decision making

    c.   Allows us to make rapid responses to people and situations

    d.   Guide a person towards rewarding outcomes, towards satisfying particular needs

e.   Could serve a self-protecting or self-defensive role

f.   Aid in denying unpleasant truths about ourselves

PO: 5.   Describe and differentiate between the Thurstone, Likert, and Guttman scales for attitude measurement.

14.   How does the researcher try to measure attitudes?   (HF p. 178)

-    By trying to measure the various components

-    Strength of feeling

-    Degree of resistance to change

-    Extent of occupation with it

-    The extent the attitude is acted upon - behavior

15.   What component of attitude is most directly measurable?   (HF p. 178)

-    The behavior

16.   What is the best known way of trying to determine attitudes?   (HF p. 178)

a.   The opinion poll

b.   Usually trys to determine the direction rather than the strength of the attitude

17.   What are the characteristics of the Thurstone, Likert, and Guttman scales for measuring attitudes?   (HF p. 178)

a.   Thurstone Scale:

-    Opinions placed in a sequence from one extreme to the other

-    Mark the opinion you agree with

-    The value gives both strength and direction

-    Not a cumulative scale

174

b.   Likert Scale:

-       A list of opinions given

-       Variable response provided for

Strongly approve ---> Strongly disapprove

1   2   3   4   5

-       Direction and magnitude is the sum of all the scores

c.   Guttman Scale:

-       A set of items measuring a single, unidirectional trait along a
        continuum of magnitude

-       A cumulative scale, implies acceptance of the lesser magnitude

18.   What is one way these attitude surveys or opinion polls may be distorted?
                                                                        (HF p. 178)
-       Due to the person's desire to be seen in a favorable light

*PO: 6.   Describe, in terms of social psychology, the difference between a collection of
        individual people and a group.*

19.   In what way is a group more than simply a collection of people?   (HF p. 179)

a.   They have shared characteristics that set them apart

-       Give them a sense of belonging

-       Shared goals, values, interests and motives

b.   Usually those who join a group feel the majority of the group share their own
        views

c.   The group has a profound and extensive effect on the attitudes, beliefs,
        interests, behavior and even goals of the individual member.

*PO: 7.   Describe some of the attitudes and behavior of individuals who belong to, and
        are being influenced by the group.*

20. What are three types of group influenced behavior which may have a detrimental effect on the individual performance with regard to safety?

(HF pp. 179-181)

a. Risk-taking

b. Loss of inhibition

c. Conformity

21. What influence does a group decision have on the element of risk-taking?

(HF p. 179)

a. It is likely to involve a greater element of risk than a decision made by an individual

b. This is contrary to the expectancy that compromise will eliminate the more risky alternatives

22. What are the reasons a group has a tendency to accept greater risk than an individual?

(HF p. 180)

a. Diffusion of responsibility

- Any adverse consequences are shared by all the members of the group

b. Dominance of risk-taker leadership

- Theory that risk-taker personalities are more likely to assume leadership roles

c. Increased familiarity

- Group discussions increases familiarity with all aspects of the topic

- Provides confidence to accept more risk

d. Social acceptance

- There is experimental support for the assumption that:

- Risk is a socially desirable value

- Most social heros are risk-takers

- Based on two suppositions:

176

1- An individual normally over-estimates their personal level of risk-taking

2- Once in a group each individual finds out their own level of risk-taking is not so high, relative to others

- As a result each one raises their own level and the group average level is shifted upward

23. What causes the loss of inhibition an individual experiences when in a group?
(HF p. 180)

- Something in the presence of others in the group weakens an individual's maintenance of socially acceptable norms of behavior

    - Party or football game behavior

    - Mob violence and gang rape

a. Probably comes from the feeling of anonymity in a group

b. Also generated from the diffusion of responsibility

c. The leadership of bad example:

    - Undesirable behavior of some may act as a model for others to follow

24. What are the factors which constitute the influence of a group on the individual to maintain conformity? (HF p. 181)

a. The larger the group holding a particular attitude the greater the pressure to conform

b. Often results in the distortion of perceptions, judgment and action

c. Can have both positive and negative applications

d. Is affected by reinforcement - rewards and punishment

25. When are group pressures to conform less effective on an individual? (HF p. 181)

a. If an individual has confirmed the validity of their own attitudes and behavior

b. When the individual has a background of success in the subject or attitude in question

177

26.   What makes a person's attitude resistant to change?   (HF p.182)

-           We tend to dispute or reject any evidence presented that appears to contradict our strongly held attitudes

27.   Which attitudes are more easily changed?   (HF p. 182)

a.   Those which are not central to a belief

b.   Those not an essential part of one's basic perception of life and living

28.   How can a person's resistance to changing their attitude be altered?   (HF p. 182)

a.   Decreased   -   Role reversal

-           Tends to result in a less intransigent position

-           Reduces the practice of only accepting those points that support our own point of view

b.   Increase   -   Elicit a personal commitment

-           Inoculation method

-           Expose people to small doses of certain undesirable attitudes and thus increase resistance to change

29.   What can be done to improve undesirable personality traits in cockpit crew members?
                                                                    (HF p. 183)

a.   It is very unlikely any training program will improve personality traits

a.   Poor personality traits must be screened out during the selection process

c.   Pilot tasks are changing from airplane driver to systems manager

-           Unfavorable personality types for the tasks must be defined

-           Screening must take into account the job as it is now and as it may become within the lifetime of the applicant

30. How effective will training be in changing the attitudes of a pilot?   (HF p. 184)

   a.   It depends on how strong the attitude is

   b.   It depends on the power of motivation used in the training

PO: 9.   *Describe the characteristics of communications designed to change attitudes.*

31. What are the functions of communications?   (HF p. 184)

   a.   Instrumental: -   Obtaining something

   b.   Informative: -   Finding out or explaining

   c.   Ritual: -   Part of a ceremony

   d.   Persuasive: -   Modification of attitude or behavior

32. What is the most significant function of communication?   (HF p. 184)

   a.   Modifying attitudes or behavior

   b.   This is of direct relevance to safety and efficiency

   c.   Everyone is concerned with this function

   d.   Usually involves a two-way process of persuasion

33. What are the three basic parts of the persuasive communication process?   (HF p. 185)

   Communicator   ---> Message   ---> Audience

34. What characteristics must the person have to be an effective user of persuasive
   communications?   (HF pp. 185-186)

   a.   Credibility: -   Expertise

                       -   Trustworthiness

        -   Message is more effective if it is known the presenter has done their
            home work

   b.   Something in common with the audience

c.   Sold on the idea personally

35.   What can be done to enhance the persuasive effectiveness of the message itself?

(HF p. 187)

a.   Show only one side of the argument

- If the audience already agrees in principle

- More educated or intelligent audience is less likely to accept

b.   Show both sides of the argument

(With a stronger case against the undesired)

- If the audience does not already agree

- Which side first?

- Primacy -   if the first argument is more influential

- Recency -   if the second argument is more persuasive

c.   Include a conclusion

- Yes - Complex issues or less bright audience

- No - Simple issues or hostile or intelligent audience

d.   High fear content is usually more effective, but must not be over done

e.   Repetition usually increases effectiveness

f.   Use of music arouses positive feelings

g.   Medium of transmission

- Single face-to-face is usually most effective

36.   How does the nature of the receiver affect the effectiveness of persuasive
communication?

(HF p. 187)

a.   Very high or very low self esteem will have the least attitude change.

b.   Sometimes a degree of distraction during the presentation seems to increase the
persuasiveness of the message.

37. What are some of the reasons that any single persuasive safety communication does not seem as effective as we might hope? (HF pp. 187-188)

    a.   Lack of direct involvement -   It will happen to the other guy

    b.   Many factors in an accident

    c.   Many safety campaigns are not as effective as they could be

    d.   A proper overall safety attitude must be developed gradually

    e.   The attitude must be constantly fostered and reinforced

    f.   For passengers:

          -    Are not available for long-term indoctrination

          -    Sometimes get a mixed message:

              -    Flying is safe
                      but
                you must listen to this "what if" message

ATTITUDES AND PERSUASION

Personality Traits:

- Deep-seated characteristics which constitute the essence of a person

- Are stable and very resistant to change

Attitudes:

- Likes and dislikes

- A learned and rather enduring tendency

- A predisposition to respond in a certain way

Beliefs:

- Not necessarily a favorable or unfavorable evaluation

- An assertion about two objects

Opinions:

- A verbal expression of an attitude or belief

ATTITUDES

- Hypothetical constructs:

    - Personality

    - Attitudes

    - Belief

    - Opinion

- Components:

    - Cognitive   -   knowledge, idea or belief about a subject

    - Affective   -   feelings held about it

    - Behavioral   -   what is said or done about it

MEASUREMENTS OF ATTITUDES

Thurstone Scale:

-   Opinions placed in a sequence from one extreme to the other

Likert Scale:

-   Variable response

    Strongly approve ---> Strongly disapprove

    1   2   3   4   5

Guttman Scale:

-   A set of items measuring a single, unidirectional trait along a continuum of magnitude

Distortion -   Due to the person's desire to be seen in a favorable light

GROUP INFLUENCES

Risk-taking

Loss of inhibition

Conformity

RISK-TAKING

a. Diffusion of responsibility

b. Dominance of risk-taker leadership

c. Increased familiarity with the problem

d. Social acceptance of risk-taking

LOSS OF INHIBITION

a. Feeling of anonymity in a group

b. Diffusion of responsibility

c. The leadership of bad example

FUNCTIONS OF COMMUNICATION

a. Instrumental: -   Obtaining something

b. Informative: -   Finding out or explaining

c. Ritual: -   Part of a ceremony

d. Persuasive: -   Modification of attitude or behavior

PERSUASIVE COMMUNICATION

Communicator ---> Message ---> Audience

PERSONAL CHARACTERISTICS
OF THE
EFFECTIVE PRESENTER

a. Credibility: - Expertise

- Trustworthiness

b. Something in common with the audience

c. Sold on the idea personally

TO ENHANCE THE MESSAGE

a.   Show only one side of the argument

b.   Show both sides of the argument

c.   Include a conclusion

d.   Repetition usually increases effectiveness

e.   Use of music arouses positive feelings

f.   Medium of transmission

WHY DOESN'T SAFETY SELL?

a. Lack of direct involvement

b. Many factors in an accident

c. Not effectively presented

d. It takes time to develop the safety attitude

e. The attitude must be constantly fostered and reinforced

f. For passengers:

   - They are not available for long-term indoctrination

   - Sometimes they get a mixed message

DAY 16   Training and Training Devices

PO: 1.   *Define the terms education, training, and skill as used in the text and differentiate between them.*

1.   What are some of the subjects included in the study of the human factors in training devices?

    a.   Definitions

    b.   Alternatives to Training

    c.   Training Principles

    d.   Training , Learning and Memory

    e.   Training Systems

    f.   Training Devices

    g.   Instructors and Classrooms

2.   How important is training to our lives in general and to aviation in particular?

(HF p. 189)

    -    We will be involved in either receiving or giving it most of our lives.

3.   How does training fit into the SHEL conceptual model?   (HF p. 189)

    a.   We expect the Liveware to possess the knowledge and the skills to fulfill its role in the system.

    b.   The preparation for specific job abilities requires involvement in the teaching process.

4.   How would you define **Education**?   (HF p. 189)

   a.   A broad-based set of   -   knowledge
                                     -    values
                                     -    attitudes
                                     -    skills

   b.   Suitable as a background upon which more specific job abilities can be
         acquired at a later stage

5.   What is **Training**?   (HF p. 189)

   a.   A process aimed at developing specific   -   skills
                                                   -    knowledge
                                                   -    attitudes

   b.   Education is seen as the precursor or foundation of training.

   c.   Certain basic educational criteria is part of staff selection.

6.   Where does the term **Instruction** fit into the training concepts?   (HF p. 190)

   -    Activities associated with either education or training

7.   What are **Skills**?   (HF p. 190)

   a.   An organized and coordinated pattern of activity

         -    Physical
         -    Social
         -    Linguistic
         -    Intellectual

   b.   Two elements: -   Deciding on a course of action

                              -    Carrying out that action

   c.   Possession of a skill in a certain activity does not imply the person can teach
         the skill in that activity

         -    Teaching is a skill in and of itself

191

8.   How would you differentiate between knowledge and skill?   (HF p. 190)

Knowledge   = Knowing <u>WHAT</u> is required to perform a task

Skill   = Knowing <u>HOW</u> to perform the task

PO: 2.   *Describe some of the alternatives to training.*

9.   What are some of the alternatives to investing in training programs?   (HF p.190)

a.   Take the task from the human altogether and assign it to a machine

1-   High performance aircraft manoeuvering - auto pilot

2-   Navigation - computer

3-   Flight management - computer

b.   Staying with the human there are three alternatives:

1-   Refine the selection process:

-    for higher education
-    for prior training
-    for special aptitude

2-   Redesign the job or job situation

-    modifying the task to fit the human capability better

3-   Spend to develop an adequate training program

-    modify the human capability to fit the task better

PO: 3.   *Define and explain the principles of training transfer, feedback, guidance, cueing, prompting, and pacing.*

10.   How would you explain the concept of **training transfer**?   (HF p. 191)

a.   What is learned in one situation can be used in another

b.   **Positive** transfer supports new learning

c.   **Negative** transfer detracts from new learning

-    Requires "unlearning" before learning

-    Standardization of equipment and procedures reduces negative training transfer

-    Sometimes difficult to unlearn the earlier practice particularly under stress

d.   Especially important in use of simulators

11.   How would you define **feedback**?   (HF p. 192)

-    The output of a system can regulate or control the input

-    The thermostat - a closed loop system

12.   How do you differentiate between the open and closed-loop feedback systems?

Refer to Fig 9.1 ----->   (HF p. 192)

a.   **Closed-loop** = feedback is present and used to regulate the input

b.   **Open-loop**   = no feedback is provided and deficiencies are difficult to detect

13.   What is the difference between intrinsic and extrinsic feedback?   (HF p. 193)

a.   **Intrinsic** Feedback

-    Available in the normal job situation

-    Native to the situation like changing visual cues

b.   **Extrinsic** Feedback

-    Additional information not available in the job situation

-    Instructor input

-    Simulator "freeze" to allow time for the student to catch up

-    Training must be designed so that the removal of extrinsic feedback will not adversely effect performance

193

14. How would you define and describe the difference between guidance, cueing and prompting? (HF p. 193)

    a. **Guidance** = is physical control of movement

        - Control movements in the flare or formation training

    b. **Cueing** = restricted to perceptual detection tasks

        - Informing a trainee when a signal is about to appear

        - Focus of attention on an event or cue

    c. **Prompting** = presenting the trainee with the correct response immediately after the stimulus

    d. All of these techniques are extrinsic and must be gradually removed as training progresses

15. What does **pacing** refer to in the training scenario? (HF p. 194)

    a. The rate at which material and experience is presented to the student

    b. Self-paced may be advantageous in the learning process

        - Programmed forms of instruction are good for this

    c. Pacing becomes more difficult as the class size increases in the traditional discussion/lecture form

PO: 4.   *Explain the stages in the learning process, how memory can be improved, and some of the handicaps to learning.*

16. What is the difference between learning and training? (HF p. 194)

    - **Learning** ---> an internal process <--- control of the process = **Training**

    - Learning <--- Student **** Instructor ---> Training

    - Success is judged by changes in performance

17. What are the three phases of learning?   (HF p. 194)

    a. Cognitive stage - Talking about the task and possible errors

                                                   (Ground school)

    b. Associative stage - Practical or practice to reduce the errors

                                                   (Dual)

    c. Autonomous stage - Perfection of performance

                                                   - Speeding up

                                                   - Improving accuracy and precision

                                                   (Solo)

18. What are the two kinds of memory and the characteristics of each? (HF pp. 194-195)

    a. Short-term memory (STM)

        - Time - Accurate recall within a few seconds

        - Capacity - Six to eight items or chunks of information

        - Processing - Limited level

        Refer to Fig 9.2 ----->   (HF p. 195)

    b. Rehearsal - Improves retention for longer time
                      - Helps transfer information to long-term memory

        Refer to Fig 9.3 ----->   (HF p. 195)

    c. Long-term memory (LTM)

        - Time - More time to accomplish than STM

        - Capacity - Storage space is no problem

        - Processing - Preparation for storage is important

                 - Coding and organization have major influence

                 - Involves time to summarize and emphasize

19. What are some of the common memory enhancement techniques?   (HF pp. 196-197)

    a.   Organization -   Mnemonics

        1-   Information put into a rhyme

        2-   Relating items to places

            -    Form a mental picture of the item and its location

            -    The more unusual the picture the better

        3-   Key-word system

            -    Replaces number to remember with consonant sounds

    b.   Over-learning:

        -    The most important concept to be built into a training program

        -    Carrying the training process beyond the minimum level of acceptance

        -    Improves recall

        -    Increases resistance to stress

    c.   Chunking:

        -    Breaking information down into chunks or familiar units

        -    The number of retainable chunks is quite constant

        -    The individual chunks can vary considerably in complexity and length

20. What effect does sleep have on memory in the learning process?   (HF p. 197)

    -    More effective to learn right before rather than right after

21. What effect does age have on memory?

    Refer to Fig 9.4 ----->   (HF p. 198)

    a.   It is difficult to generalize because of the differences between individuals

    b.   Usually not very dramatic until reaching the sixties

c.   Deterioration is selective - some memory holds up better than others

d.   Depression is a condition damaging to memory

e.   Mental function is like physical it deteriorates with disuse

22.   What are the things which can become a handicap to learning?   (HFp p. 198-199)

   a.   Anxiety:

      -        Fear of failing

      -        Fear of flying

   b.   Stress:

      -        Social or domestic

   c.   Low motivation:

      -        When career prospects are determined by the outcome of training

      -        If there are feelings of unfair treatment

      -        Results from poor quality of instruction or instructional materials

      -        Reading inefficiency - very common

      -        Poor student learning technique

      -        Inadequate communication

PO: 5.   *Describe each of the various training systems, and explain the advantages and disadvantages of each.*

23.   What should be considered when selecting any given training system?   (HF p. 199)

   a.   Individual temperament and cognitive style (personal preference)

   b.   Other criteria which determine the optimum training method

   c.   Practical constraints of the organization

24.	What are the advantages and disadvantages of the **lecture**?	(HF p. 200)

    a.	Advantages:

        -	Introduction or give general background information

        -	Briefing or summary of a program

        -	No audience size limitation

    b.	Disadvantages:

        -	Not very effective in giving job skills

        -	Not very effective in changing attitudes

        -	Encourages passivity

        -	Hard to hold interest

        -	Very low feedback to the lecturer

25.	What are the pluses and minuses of the **lesson**?	(HF p. 200)

    a.	Advantages:

        -	Structured for participation

        -	Flexibility to modify pacing and presentation

        -	Permits instruction of theory in the classroom or skills in the laboratory

    b.	Disadvantages:

        -	Limited size

        -	Slower learner paced

26.	What are the good and bad points of **discussions**?	(HF p. 200)

    -	Question / answer session guided by the instructor

    -	Group discussion of all participants

a. Advantages:

- Participation is strongly encouraged

- High interest level

- Usually high learning accomplished

b. Disadvantages:

- Cannot be highly structured

- Requires skilled instructor to guide to objectives

- Limited size of participants

- Time expensive

27. How does the **tutorial** system work for efficiency?   (HF p. 200)

- Instructor works directly with only one student

a. Advantages:

- Good for complex skills

- Good for safety where training risks are involved

- Good for individual remedial training

- Excellent feedback and participation

b. Disadvantages:

- Very expensive

- Student does not benefit from observing other student performance

28. What are the advantages and disadvantages of **audio-visual** methods?   (HF p. 201)

- Uses tape-slide or video-tape or film presentations

- Participation can be provided before or after the presentation

199

a. Advantages:

- Prepared to maximize both audio and visual aspects

- Very flexible for location and audience size

- Skill required to make the presentation doesn't have to be available at the presentation

- Can contain a great amount of information

- Repetition of presentation is simple

b. Disadvantages:

- Low participation

- Modification and updating of programs involve skilled technical work

29. What are the values and drawbacks of **programmed instruction**?   (HF p. 201)

- Based on student signaling

a. Advantages:

- Usually a closed-loop system

- Minimized training costs

- Can be audio-visual with advanced computers

b. Disadvantages:

- If fully automated - no way for student questions or discussion

- Takes considerable expertise and work to develop

30. How would you describe the pros and cons of a system using **computer-based training**?   (HF p. 201)

- Also known as Computer Assisted Instruction (CAI)

a. Advantages:

- Programs can be highly structured

-   Have great flexibility

-   Can be linked with personnel records

-   Can incorporate testing

-   Good self pacing options

-   Participation built in to questions and touch sensitive screens

-   Provide remote capability

b.   Disadvantages:

-   Significant expertise required to develop

-   Initial production cost high

PO: 6.   *Describe what is entailed in the systems approach to training.*

31.   What is meant by using the systems approach in training?   (HF p. 202)

a.   All the components and their inter-relationships are studied before the training program is put into effect

Refer to Fig 9.5 ----->   (HF p. 202)

b.   Starts with job analysis

c.   End objectives are developed

d.   Trainee selection criteria is established

e.   Training course development

f.   Various feedback loops and validations are built in

PO: 7.   *Explain the difference between training aids and training equipment, and what considerations must be made for their procurement.*

32.   What are the two major classes of training devices?   (HF p. 203)

a.   Training aids -   Devices used by an instructor to help present the subject

-   Slides or view-graphs

- Chalkboards etc.

    b.   Training equipment -   Provide for some active participation and practice

- Touch sensitive control CRT

- Flight simulator

- The way the hardware is used rather than the nature of the hardware itself

33.   What things must be considered when procuring training devices?   (HF p. 203)

    a.   Reliability

    b.   Maintainability

    c.   Simplicity

    d.   Efficient learning of the specific task for which it is required

    e.   Should allow for practice

    f.   Should provide feedback

PO: 8.   *Describe the concept of fidelity as it relates to training devices and training efficiency.*

34.   What are the two powerful incentives for the development of simulators? (HF p. 203)

    a.   Provide practical training in an equivalent environment

- at a lower cost

- with less risk

    b.   Provide sufficient fidelity to

- Obtain authorization from state certifying authorities for the simulators to be used as measures of the human proficiency

35.   What does the degree of fidelity refer to in describing a simulator's performance?

(HF p. 204)

- The accuracy or faithfulness with which a simulator reflects the real task

36. How much fidelity is required for effective training?   (HF pp. 204-205)

    a.   It depends on the nature of the training

    b.   It must be applied to each training situation, taking into account the nature of the specific task being simulated

    c.   The degree of fidelity may have little impact

    d.   Simulator features like freeze may enhance training but depart from true fidelity

    e.   Lack of certain kinds of fidelity can have negative training transfer

        -     Flight control feel or stability

    f.   Fidelity costs money

    Refer to Fig 9.7 ----->   (HF p. 205)

37. What SHEL interfaces should have high or low fidelity?   (HF pp. 205-206)

    a.   L - H -   controls and displays -   high

    b.   L - E -   Noise, temperature and vibration - low

38. What training situations may require high fidelity?   (HF p. 206)

    a.   When a trainee must learn to make discriminations

    b.   When responses required are either difficult to make or

    c.   When responses are very critical to the operation

39. When would a low fidelity device be preferable?   (HF p. 206)

    a.   When only procedures are being learned

    b.   When starting training

(next page)

40. What does psychological fidelity depend upon? (HF p. 206)

    a.   The perception of the training device by the individual student

    b.   It is not necessarily dependent upon either equipment or environment fidelity

TRAINING AND TRAINING DEVICES

- Definitions

- Alternatives to Training

- Training Principles

- Training , Learning and Memory

- Training Systems

- Training Devices

- Instructors and Classrooms

DEFINITIONS

Education:

- A broad-based set of
    - knowledge
    - values
    - attitudes
    - skills

- Suitable as a background upon which more specific job abilities can be acquired at a later stage

Training:

- A process aimed at developing specific
    - skills
    - knowledge
    - attitudes

Instruction:

- Activities associated with either education or training

DEFINITIONS (cont)

Skills:

- An organized and coordinated pattern of activity

    - Physical
    - Social
    - Linguistic
    - Intellectual

- Two elements:

    - Deciding on a course of action

    - Carrying out that action

Knowledge

= Knowing <u>WHAT</u> is required to perform a task

Skill

= Knowing <u>HOW</u> to perform the task

ALTERNATIVES TO TRAINING

- Assign tasks to a machine

- Refine the selection process

- Redesign the job

- Spend to develop a adequate training program

TRAINING PRINCIPLES

Training Transfer

- Positive transfer

- Negative transfer

FEEDBACK

Feed-back Control

- The output of a system can regulate or control the input

- Open-loop

- Closed-loop

Intrinsic Feedback

- Available in the normal job situation

Extrinsic Feedback

- Additional information not available in the job situation

ADDITIONAL TERMS

Guidance   =   Physical control of movement

Cueing   =   Restricted to perceptual detection
tasks

-    focus of attention on an event or
cue

Prompting   =   Presenting the trainee with the correct
response immediately after the
stimulus

LEARNING & TRAINING

Learning --->

Internal process

<--- Control of the process = Training

Learning <--- Student ** Instructor ---> Training

PHASES OF LEARNING

- Cognitive stage - Talking about the task

 (Ground school)

- Associative stage - Practical or practice

 (Dual)

- Autonomous stage - Perfection of performance

 (Solo)

MEMORY

Short-term memory (STM)

- Time - Short input short retention

- Capacity - Six to eight items

- Processing - Limited level

Long-term memory (LTM)

- Time - Long input long retention

- Capacity - Storage space is no problem

- Processing - Significant coding and organization

MEMORY ENHANCING

a. Organization

b. Over-learning

c. Chunking

LEARNING HANDICAPS

a. Anxiety: - Fear of failure

- Fear of flying

b. Stress: - Domestic or social

c. Low motivation:

- Career prospects

- Feelings of unfair treatment

- Poor quality of instruction

- Reading inefficiency

- Poor student learning technique

- Inadequate communication

TRAINING SYSTEMS

- Lectures

- Lessons

- Discussions

- Tutorial

- Programmed Instruction

- Computer-based training

TRAINING DEVICES

a. Training Aids   -   Devices used by an instructor to help present the subject

b. Training Equipment   -   Provide for some active participation and practice

INCENTIVES FOR SIMULATION

a.   Provide practical training in an equivalent environment

      -    at a lower cost

      -    with less risk

b.   Provide sufficient fidelity to

      -    obtain authorization from state certifying authorities for the simulators to be used as measures of the human proficiency

USE HIGH FIDELITY

a.   When a trainee must learn to make discriminations

b.   When responses required are either difficult to make

c.   When responses are very critical to the operation

USE LOW FIDELITY

a.   When only procedures are being learned

b.   When starting training

HUMAN FACTORS IN AVIATION SAFETY
(SF 320)
LECTURE NOTES

DAY 17   Documentation

PO: 1.   *Describe how some aspects of language can effect the efficiency of written documentation.*

1.   What are some reasons it is important to have proper and correct documentation in the aviation industry?   (HF p. 209)

   a.   Errors in documentation can be very expensive.

   b.   Poor documentation wastes time and lowers efficiency.

   c.   Business potential is reduced from poorly written materials.

   d.   Safety may be compromised as a result of poor documentation.

   e.   Consider the vast amount of paper work required to get an aircraft airborne.

      -    A fundamental law of aerodynamics:

      Weight of the paper must => Weight of the aircraft

2.   What are some of the requirements any technical writer must meet to produce an effective document?   (HF p. 209)

   a.   Skill in presenting graphic material

   b.   Writing to ensure accuracy and comprehension

   c.   Avoid ambiguity

   d.   Maintain motivation

   e.   Responsive to practical needs

   f.   Understand how the document is to be read and used

3.   What part of documentation should display consistency?   (HF p. 210)

   a.   Spacing

b.   Tables

c.   Illustration

d.   Style of headings

e.   References´

f.   Quotations

g.   Numbering

h.   Dates

i.   Foreign words

4.   What are the three basic aspects which require human factors optimization?
(HF p. 210)

a.   Written Language

b.   Printing

c.   Layout

5.   What is basic about the choice of language for written documentation?   (HF p. 210)

a.   The language used must be that of the reader

b.   Even English expressions between the U.S and the U.K. have different meanings

c.   Special care must be taken when writing a language different from our own

6.   What are the primary problems which must be faced in good technical writing?
(HF p. 211)

a.   Conceptual

b.   Procedural

7.   Who is better qualified than the linguist or literary scholar to use the language and words of machines?
(HF p. 212)

-    The human factors specialist

8. How should writing skills be focused for communicating effectively with passengers?

(HF p. 212)

    a. Understand the audience

    b. Optimize for comprehension

9. What are some of the general rules directed toward optimizing comprehension?

(HF p. 213)

    a. Shorter and more familiar words are better

    b. Shorter sentences are better

       - Less than 20 words generally -- good

    c. Don't cut out critical words for brevity

    d. Only use appropriate abbreviations

       - Too many abbreviations can be very difficult to read

10. What is "jargon," and what guidance is given for its use in technical documentation?

(HF p. 214)

    a. Refers to any specialized vocabulary

    b. It seems to be an unavoidable component of technical writing

    c. It serves a useful purpose if not employed excessively

    d. It is hard on the reader who is new to the material

11. How can you evaluate the reading level of technical material?

    - By taking an average word and average sentence length in a given sample of text

        Refer to Appendix 1.20 ----- >   (HF p. 348)

PO: 2.   *Describe how various aspects of printing can effect the efficiency of written documentation.*

12. What, in general, is being considered when evaluating printing for documentation efficiency?

(HF p. 214)

a. Typography - the form of letters

b. Printing and layout have a significant effect on the comprehension of the written material

13. How is print size considered in documentation? (HF p. 215)

a. Uses a point system of measuring

b. Letters for most printing should not be less than 10 points

- = 3.5 mm from top of high letters to bottom of low ones

c. Older people prefer 11 or 12 point type size

14. What can be done to increase legibility of smaller print? (HF p. 215)

a. Increase the space between the lines of type

b. This is called leading

c. For typing 1 1/2 space is better than single space

15. What are the two general distinctions used for type-face styles? (HF p. 215)

a. Sanserif - without decorative strokes

b. Serif - with decorative strokes

16. Which group of type-faces is best for documentation? (HF p. 215)

a. Serif is better in some tests for long passage comprehension

b. Sanserif better for good legibility and placards

17. How does upper or lower case affect readability? (HF p. 215)

a. Usually words set in UPPER CASE (capitals) are less readable than lower case

b. Lower case should always be used except to apply normal capitalization rules

c. Use of **bold face** for headings is better than UPPER CASE

d.   In typing <u>underline</u> is better than UPPER CASE

18.   What effect do *italics* have on readability?   (HF p. 216)

-      *Italics are harder to read* than normal print

PO: 4.   Describe how various aspects of layout can effect the efficiency of written documentation.

19.   What is going to be the major factor for determining the layout of a document?

(HF 216)

a.   The size and shape of the document

b.   Size and shape may be largely determined by the way a document is intended to be used

c.   Functional layout may determine shape or size -   checklist

20.   What type of documentation would use single vs double column layout?   (HF p. 217)

a.   Single   -   Presentation of complex instructional material

-      When a large number of tables or diagrams

b.   Double   -   Easier for speed and scanning -   Newspaper

21.   How does justification effect the readability of text?   (HF p. 217)

a.   Justification = right edge made even

b.   Justified text aids reading - no evidence

22.   How should spacing for paragraphs and other elements be used?   (HF p. 217)

a.   Short text paragraphs should use spacing rather than indentation

b.   Spacing is used where distinctive grouping of information aids accurate and rapid access and comprehension

c.   It is important when trying to create a favorable visual impression for interest and motivation

23.   What is an essential requirement before the mass publication of a technical document?
(HF p. 217)

a.   The initial version must be tested to insure the point being made is understood

b.   It is normal to expect some revision of all technical documents

PO: 5.   Describe the purposes for using checklists in the cockpit of transport aircraft, and some of the ways they are used.

24.   What are some of the purposes for a checklist in the cockpit of a transport aircraft?
(HF pp. 218-219)

a.   Reduces the risk of missing items

b.   Helps keep technically significant sequences in order

c.   Allows assigning as many items as possible to low workload periods

d.   Designed for convenient eye and arm movement

e.   Provides for optimum distribution of tasks

f.   Encourages the team approach to cockpit tasks

g.   Ensures operational compliance

25.   How are checklists used in the cockpits of transport aircraft?   (HF p. 220)

a.   Most airlines use the hand-held-card type

b.   "Do list"   -   Items done as read from the checklist

c.   "Verification list"   -   visual and verbal verification that the item is done

PO: 6.   Describe some of the human factors principles used in effective design of technical documents.

26.   How important are page numbering, indexing and a table of contents for technical documents?
(HF p. 225)

a.   Page numbering on the left results in fewer errors

b.   Indexing and an optimum table of contents are essential for maintenance and operating manuals

c.   Numbering of paragraphs is better than lettering

d.   Color coding is useful in distinguishing between certain sections (Be careful to consider lighting)

27.   What is the recent, significant development for helping to present technical information?   (HF p. 227)

-      Use of illustrations

28.   What are the functions of illustrations in technical documentation?   (HF p. 227)

a.   Motivate the reader

b.   Help in recall from long-term memory

c.   Aid in explanation

d.   Help avoid jargon

29.   What has contributed to the 50-80 percent human error rate causing system failures?   (HF p. 227)

-      Inadequacy in manuals, handbooks and written instructions

PO: 7.   *Describe the principles used to enhance the effectiveness of questionnaires and forms as written documentation.*

30.   What can questionnaires or surveys be used for?   (HF p. 229)

a.   To discover peoples opinions or attitudes

b.   To acquire data not directly available from experimental studies

31.   What is the value of the survey dependent upon?   (HF p. 229)

-      The design of the survey program

32.   How should the survey program be designed to be most effective?

Refer to Fig 10.3 ----->   (HF p. 230)
         (HF pp. 229-231)

a.   Identify clearly the problem

b.   Define the information needed

c.   Review the relevant literature

d.   Establish who is to give the answers

e.   Select an appropriate method of collection

    -    Personal interview is preferred

f.   Design the survey questions and make a trial survey

g.   Analyze the results

33.   What is the most important objective in the design of the survey questions?

(HF p. 231)

a.   Prevention of bias or distortion of answers through prejudice, suggestion or other extraneous influences

    Common sources of bias are:

    -    Fear

    -    Social pressures to conform

    -    Desire to please the experimenter

34.   What are some of the types of questions that can be used in a questionnaire?

(HF pp. 231-233)

a.   Factual

    -    Factual answers can usually be checked for validity

    -    Problems: definition, comprehension, memory and bias

b.   Non-factual

    -    More difficult and impossible to validate

c.   Open-ended

    -    Usually provides information with little bias

-       Time consuming and difficult to process

d.   Closed-ended

-       Greater risk for bias

-       Quick to answer and easier to process

e.   Rating scales

-       Quick to handle and versatile

-       There is a risk of bias

-       Interpretation may be difficult

-       A graphic form is a simple 10 cm line with two extremes

-       Common bias is the reluctance to select extremes

35.   How does the "order effect" influence the information derived from a survey?

(HF p. 233)

a.   The first mentioned alternative may be favored

b.   The order of placement relative to other questions may influence the response

36.   How can wording affect the outcome of the information received on a survey?

(HF p. 234)

a.   Sensitive questions require special attention

b.   Loaded words or loaded questions can slant the response

37.   What are the three "Ls" of form design?   (HF p. 236)

a.   Layout

b.   Logic

c.   Language

PO: 8.   *Explain how the application of human factors can be used to improve the documentation found on maps and aeronautical charts.*

38.   What is the difference between a map and a chart?   (HF p. 237)

     a.   Maps   -   Generally topographical

                 -    Have many purposes

                 -    More permanent

     b.   Charts   -   Nautical or aeronautical

                 -    Used only for navigation

                 -    Subject to periodic revision

39.   How has a systematic and professional application of human factors been applied to the field of chart documentation?   (HF p. 237)

     a.   It has been notably lacking

     b.   Generally the needs of the user have not had sufficient influence

40.   What are the two basic reasons for neglecting an adequate consideration of human factors in chart making?   (HF p. 237)

     a.   Long tradition in the hands of specialized cartographers

     b.   Research into human factors of charts is complicated and expensive

41.   What things need to be done to improve the human factors applications to charts and maps?   (HF pp. 238-239)

     a.   Detailed task analysis

                 -    How the charts are used

                 -    What information is required to be displayed

                 -    Adapting to the different operational uses

     b.   Optimized for use on the flight deck

                 -    Size vs working space

                 -    Viewing distance and angle

-       Accounting for vibration and turbulence

-       Considering color brightness and various illumination

-       Subject to rough handling, heat, moisture or direct sunlight

42.   What was used in the 1940s by ICAO as a basis for the standardization of the symbols which appears on aeronautical charts?   (HF p. 239)

a.   What was in common use at the time

b.   Human factors research and definition was not applied

Documentation

Skills for Effective Documentation

a. Skill in presenting graphic material

b. Writing to ensure accuracy and comprehension

c. Avoiding ambiguity

d. Maintaining motivation

e. Responsive to practical needs

f. Understanding the documents used

Documentation Consistency

    a.   Spacing

    b.   Tables

    c.   Illustration

    d.   Style of headings

    e.   References

    f.   Quotations

    g.   Numbering

Human Factors Optimization Requirements

    a.   Written Language

    b.   Printing

    c.   Layout

Optimizing for Comprehension

a.   Shorter and more familiar words are better

b.   Shorter sentences are better

-   Less than 20 words generally -- good

c.   Don't cut out critical words for brevity

d.   Only use appropriate abbreviations

Print Size in Documentation

a.   Uses a point system of measuring letters

b.   Should not be less than 10 points

c.   Older people prefer 11 or 12 point type size

Sanserif   -   without decorative strokes

            -   Better for good legibility

Serif   -   **with decorative strokes**

            -   Better for long passage comprehension

UPPER CASE   vs   lower case

UPPER CASE (CAPITALS) ARE LESS READABLE
than lower case

UPPER CASE vs   **Bold Face**

bold face for headings is better than UPPER CASE

In typing <u>underline</u> is better than UPPER CASE

Italics   vs   Normal

— Italics are harder to read
than normal print

Purposes for Checklists in the Cockpit

a.   Reduces the risk of missing items

b.   Helps keep proper sequence

c.   Allows assigning for workload conditions

d.   Convenient eye and arm movement

e.   Optimum distribution of tasks

f.   Encourages the team approach

g.   Ensures operational compliance

Cockpit use of Checklists

a.   Hand-held-card type

b.   "Do list"

c.   "Verification list"

Illustrations

a. Motivate the reader

b. Help in recall from long-term memory

c. Aid in explanation

d. Help avoid jargon

Questionnaire and Survey
Question Types

a. Factual

b. Non-factual

c. Open-ended

d. Closed-ended

e. Rating scales

Map or Chart

Maps - Generally topographical

- Have many purposes

- More permanent

Charts - Nautical or aeronautical

- Used only for navigation

- Subject to periodic revision

Improvements for Charts

a. Detailed task analysis

b. Optimized for use on the flight deck

HUMAN FACTORS IN AVIATION SAFETY
(SF 320)
LECTURE NOTES

DAY 19   Displays

PO: 1.   Describe the historical development of human factors in cockpit displays and controls.

1.   What are the main sources of pressure for the developments in displays and controls in modern transport aircraft?

(HF p. 241)

    a.   Safety

        -        Causal factors of poorly designed controls

        -        More displays needed for the higher performance

        -        More chance for human factors errors

    b.   Economics

        -        Larger aircraft = more victims per crash   = More litigation

2.   What were some of the earliest flight instrument displays?

(HF p. 241)

    a.   Barometer

    b.   Yaw string - slip indicator

    c.   Engine RPM indicator

3.   What were some of the major historical milestones in the development of cockpit displays?   (HF p. 241)

    a.   A usable gyroscope for an artificial horizon   (1920s - 1930s)

    b.   Serious attention given to the layout of the cockpit instruments
        (1930s)   (HF p. 243)

    c.   Electronics and servo-driven instruments   (1950s)   (HF p. 241)

        -        The sensor could be remotely located

d.   Use of the cathode ray tube (CRT) for primary flight information has had major impact.   (mid 1960s)

4.   What are the three necessary pillars upon which the progress of cockpit displays has been built?   (HF p. 242)

    a.   Avionics or aviation electronics

    b.   Ergonomics or human factors

    c.   Operational input

5.   What essential pillar for proper development of displays has been frequently neglected?   (HF p. 242)

    a.   Ergonomics or human factors

    b.   Getting the information to the cockpit without regard for the way the information was sensed or processed by the crew

    c.   The human was adapted to the display rather than the display being designed for the human

6.   What were some important early human factors studies that lead the way to better design of cockpit displays?   (HF p. 242)

    a.   Fitts, Jones and Grether studies:

        -      At Aero-Medical Laboratory in Dayton

        -      A more scientific approach to display design

    b.   Three pointer altimeter error study

    c.   Study of the effect of dial shape on legibility

    d.   Scale reading accuracy study

    e.   Numerous studies on the human factors of CRTs or visual display terminals (VDTs)

7.   What individuals formed the vital bridge between operations in air transports and the manufacturing industry?   (HF p. 242)

a.	Airline development pilots communicating with manufacturers

-	The means of translating the experience accumulated in training and line operations into the language which can be used by the equipment designers and manufacturers

b.	SAE and IATA have increasingly taken over the input responsibility

*PO: 2.	Explain the SHEL interface for displays and controls and the associated human factors problems.*

8.	How do the cockpit displays differ from the cockpit controls in the SHEL model?

(HF p. 243)

a.	Displays	=	Hardware ---> Liveware

-	Transfer information from hardware to liveware

b.	Controls	=	Liveware ---> Hardware

-	Transfer information and energy from liveware to hardware

9.	To what human senses are displays designed to transfer information?

(HF p. 243)

a.	Visual

b.	Aural

c.	Tactile

10.	What is the purpose of a display?	(HF p. 243)

-	Rapid and accurate transfer of information from the source to the brain of the crew member for processing

11.	Where does the first major human factors problem arise in the mission of display technology?

(HF p. 243)

a.	Human sensory capacity is enormous

b.	Human information transmission rate is very limited

c.	The information must be filtered, stored and processed

d.   Display design must not only present the information, but present it in a way to help the brain process it

PO: 3.   Describe and give examples of some classifications and the general design aspects of visual displays.

12.   What are the various classifications of visual displays?   (HF p. 244)

    a.   Dynamic   -   change through time

        -    Attitude indicator

        -    Altimeter

    b.   Static   -   unchanged over time

        -    Placards and signs

    c.   Quantitative   -   measurements of amounts

        -    Altimeter

        -    Heading indicator

    d.   Qualitative   -   information on rates or direction

        -    Turn coordinator

        -    Vertical Speed Indicator

    e.   System status   -   advise on a system's condition

        -    Gear indicator light

        -    Flap indicator

    f.   Warning/Caution   -   alert to malfunctions or problems

        -    Fire light

    g.   Flight Displays:

        -    Command   -   Flight director

        -    Situation   -   HSI

	-	Predictor	-	Track "noodle"

h. Inside-out - Reflects the look from inside the aircraft

- Fixed aircraft and moving background

i. Outside-in - Reflects the look from outside the aircraft

- Moving symbol and a fixed background

13. What are some general considerations for the design of cockpit displays?

(HF p. 244-246)

a. How it is used

b. What circumstances

c. Who will use it

d. Need for light and visibility

e. Angle of display - Parallax

f. Viewing distance - About 71 - 78 cm from design eye position

g. Standby mode - Clearly annunciated

14. How are visual different from aural displays with regard to their directional use in the cockpit? (HF p. 245)

a. Aural --> usually omnidirectional

b. Visual --> usually not omnidirectional

15. How would you differentiate between the digital vs the analog visual display systems?

(HF p. 245)

a. Analog --> Better for direction or trend

b. Digital --> Better for accuracy

PO: 4. *Describe some of the factors associated with the visual display of alphanumerics.*

16. Which way should the mechanical drum rotate to display numbers? (HF p. 246)

241

a. The drum should rotate down to display the next higher number

b. The system is not yet fully standardized

17. What is happening to the mechanical and electro-mechanical alphanumeric displays recently?

(HF p. 246)

\- Replaced by electronic displays

18. What human factors concerns must be addressed in electronic displays?   (HF 247)

a. Brightness   -   Sunlight or shadow use

b. Color

c. Flicker

d. Ambiguity   -   From partial failure

e. Electroluminescence (EL)

f. Incandescent lamps

g. Liquid-crystal displays (LED)

h. Light-emitting diodes

i. Gas-discharge plasma

j. Cathode-ray tube (CRT)

k. Each display must be evaluated in the light of the technology used, the specific application and the environment.

PO: 5.   *Explain the different types of dial markings and presentations, and the considerations for their use.*

19. What are the three basic types of dynamic displays?   (HF p. 247)

a. Fixed scales   with   Moving pointers

\-   Altimeter

b.   Moving scales   with   Fixed pointers

-        Directional gyro or heading indicator

c.   Digital read-out

-        DME indicator

20.   What are some of the basic human factors principles to consider in designing display scales?

(HF pp. 247-248)

a.   Avoid varying progression if possible

b.   Single unit progression is best

c.   Eliminate decimal points

d.   Graduation base usually on outside with major markers extended inward

e.   Sunburst design = graduation base on outside

f.   Pointer tip --> just touch the tip of small graduations

21.   What are the factors which determine the size of the display?   (HF pp. 248-249)

a.   Lighting

b.   Vibration

c.   Non-optimum viewing angle

d.   Full range of users

e.   Importance of use

PO: 6.   Describe some of the applications, advantages, and problems presented for cockpit displays by the use of cathode ray tubes (CRTs).

22.   Why is the use of CRTs in cockpit displays considered such a milestone? (HF p. 249)

a.   Release from many of the constraints of earlier electro-mechanical displays

b.   Permits integration of displays

c.   More effective utilization of high priority panel space

d.   More flexibility

23.   What are the general applications of CRTs as cockpit displays?   (HF pp. 249-250)

a.   Flight instrument displays

b.   Systems information

c.   Flight management systems (FMS)

24.   How are CRTs applied in the Airbus and the Boeing 757/767?   (HF p. 250)

a.   Electronic Attitude Director Indicator   (EADI)

-    Boeing

b.   Engine Indicating and Crew Alerting System   (EICAS)

-    Boeing

c.   Primary Flight Display   (PFD)

-    Airbus

d.   Electronic Centralized Aircraft Monitoring   (ECAM)

-    Airbus

-    Two CRTs: -   Warning display
                    -    Systems display

25.   What are some of the challenges to consider with the use of CRTs as flight deck displays?   (HF pp. 250-251)

a.   Brightness and brightness contrast

b.   Monochrome or color

c.   Possible fatigue effect from long periods of monitoring

d.   Software aspects of switching and time-sharing

-    What should appear where and when

e.   Sources of discomfort and annoyance:

-     Visual discomfort and fatigue
-     Blurred vision
-     Headaches and nausea
-     Cataracts
-     Muscular disorders

26.   What are some of the advantages and disadvantages of using the developing "flat panel" displays?   (HF p. 252)

a.   Offer considerable economy in weight and space

b.   Require considerably less cooling

c.   Off-angle viewing is difficult

PO: 7.   Describe some of the advantages and difficulties with the development and use of the head-up display (HUD).

27.   What area of commercial aviation is the most promising for the use of the head-up display (HUD)?   (HF p. 252)

a.   In the transition from instrument flight approach to visual landing

b.   Only 2% of landings are made in weather worse than 600/1 1/2

-     Yet these circumstances have more than 50% of the fatal accidents

28.   What are some of the applications of the HUD other than low visibility approaches?   (HF p. 253)

a.   Takeoff and climb

b.   Cruise

c.   Good visibility approaches

d.   Automatic landing monitoring

e.   Roll-out guidance

f.   Windshear protection

29. How does a HUD work? (HF p. 253)

    a.   Display is projected on a combining glass ahead of the pilot's eyes

    b.   The combining glass allows about 90% of the outside light to pass through

30. Which part of HUD technology has had the most development work? (HF p. 254)

    -    The symbology used

PO: 8.   *Differentiate between the fail-passive and fail-operational design concepts used in automatic flight control systems.*

31. What is the difference between a fail-passive and a fail- operational automatic landing system? (HF p. 254)

    a.   Fail-passive:

        -    No significant aircraft deviation

        -    No out-of-trim condition

        -    No nonapparent control problems

        -    Autopilot simply hands over the aircraft in a steady condition

        -    Should not be used in conditions below Category II without at least a HUD to assist the pilot taking over

    b.   Fail-operational:

        -    Provides sufficient redundancy to maintain capability to touchdown or roll-out

        -    One automatic system takes over another in event of a failure

32. What are the primary reasons for developing and installing the HUD in aircraft flying in bad weather? (HF p. 254)

    a.   ALPA:

        -    Raise the level of safety within current landing minimums

    b.   Airlines:

- Improve schedule regularity and reduce costs of diversions

- Allowing landing in lower visibility than otherwise would be permitted

PO: 9.   *Describe the Warning, alerting, and advisory systems and explain the differences between them.*

33.   What seems to be the trend in the development of warning systems?   (HF p. 255)

- A proliferation of systems

34.   To what kinds of things are the warning/advisory systems designed to draw the pilot's attention?   (HF p. 255)

a.   Fire

b.   Takeoff configuration

c.   Landing configuration

d.   Stabilizer trim

e.   Overspeed

f.   Altitude

g.   Autopilot disconnect

h.   Evacuation

i.   Ground proximity

j.   Decision height

k.   Cabin call

35.   What is the general concept established by SAE in 1980 with regard to warning systems?   (HF p. 255)

a.   An attention-getting sound supplemented by

b.   Voice alert messages and

c.   Discrete visual supplementary information located on a centralized panel

d.   Intended for emergency, abnormal and advisory conditions of alert

36.   What are the three fundamental objectives for application in the design of all flight deck warning systems?

Refer to Fig 11.3 ---->   (HF p. 256)

a.   Alert

b.   Report

c.   Guide

37.   What is paramount in the design of warning systems?   (HF p. 256)

a.   System reliability for credibility

b.   Excessive appearance of an alerting signal will reduce response to it

38.   What are the four functional classes of alerting systems?   (HF p. 257)

a.   Performance or departure from safe flight profiles

b.   Aircraft configuration

c.   Status of aircraft systems

d.   Related to communications

39.   How are the alerting signals grouped for priority?   (HF p. 257)

-     By using four urgency conditions:

1-   **Emergency Condition**

      -     Warning   -   Red   -   Immediate action

2-   **Abnormal Condition**

      -     Caution   -   Amber   -   Pending emergency

248

3-   **Advise**

-       Some crew action may become necessary

4-   **Alerts**

-       Provide information

40.   What are the advantages and disadvantages of the synthesized voice?

(HF p. 257)

a.   Advantages:

-       Omnidirectional

-       Attracts attention whenever the crew member's attention happens to be focused

-       Contains information usually only on visual displays

b.   Disadvantages:

-       Can be easily interfered with

-       Can interfere with other voice transmissions

41.   What is one of the essentials for cockpit displays with regard to the reliability of a display system in event of system failure?

Refer to Fig 11.4 ----->   (HF p. 258)
                                                        (HF p. 259)

a.   The display must not present unreliable information in a manner that it can be used or

b .   The information must be totally removed

DISPLAYS AND CONTROLS

Safety

Development
of
Advanced Displays & Controls

Economics

Major Historical Milestones:

- Gyroscope

- Servo-driven Instruments

- Cathode Ray Tube (CRT)

- Fitts, Jones and Grether studies

    - At Aero-Medical Laboratory in Dayton

    - A more scientific approach to display design

- Airline development pilots communicating with manufacturers

Displays   =   Hardware ---> Liveware

Controls   =   Hardware <--- Liveware

DISPLAYS
-   Purpose:

    -   Present information

    -   Present it in a way to help the brain process

Human Factors of Display Design

Enormous   Limited

Human   filtered

Sensory   stored   Brain
                                           Transmission
                                           rate

Capacity   processed

DISPLAY CLASSIFICATIONS

- Dynamic

- Static

- Quantitative

- Qualitative

- Advise on status

- Warning or Caution

- Command

- Predictor

- Situation

    - Outside-in:

    --> Moving symbol ----- Fixed background

    - Inside-out:

    --> Fixed symbol ----- Moving background

DISPLAY DESIGN

Considerations:

- How it is used

- What circumstances

- Who will use it

- Angle of display   -   Parallax

- Viewing distance   -   About 71 - 78 cm
                              from design eye
                              position

- Standby mode   -   Clearly annunciated

DISPLAY CLASSIFICATIONS (cont)

- Aural   -- > usually omnidirectional

- Visual   -- > usually not omnidirectional

- Analog -- > Better for direction or trend

- Digital -- > Better for accuracy

DIAL MARKINGS

Three basic kinds of dynamic displays

   1-   Fixed scales   with   Moving pointers

   2-   Moving scales   with   Fixed pointers

   3-   Digital read-out

DISPLAY TYPES

- Vertical

- Horizontal

- Semi-circular

- Circular

- Open window

DISPLAY SCALE
DESIGN CONSIDERATIONS

-   Avoid varying progression if possible

-   Single unit progression is best

-   Eliminate decimal points

-   Graduation base usually on outside with major markers extended inward

-   Sunburst design = graduation base on outside

-   Pointer tip --> just touch the tip of small graduations

DISPLAY SCALE
DESIGN CONSIDERATIONS
(cont)

- Size determining factors:

    - lighting

    - vibration

    - non-optimum viewing angle

    - full range of users

CATHODE-RAY TUBE
(CRT) DISPLAYS

- Flight instrument displays

- Systems information

- Flight management systems (FMS)

CATHODE-RAY TUBE
(CRT) DISPLAYS

Electronic Attitude Director Indicator (EADI)

- Boeing

Engine Indicating and Crew Alerting System (EICAS)

- Boeing

Primary Flight Display (PFD)

- Airbus

Electronic Centralized Aircraft Monitoring (ECAM)

- Airbus

- Two CRTs: - Warning display
  - Systems display

HEAD-UP DISPLAY
(HUD)

Instrument Approach ------------> Visual Landing

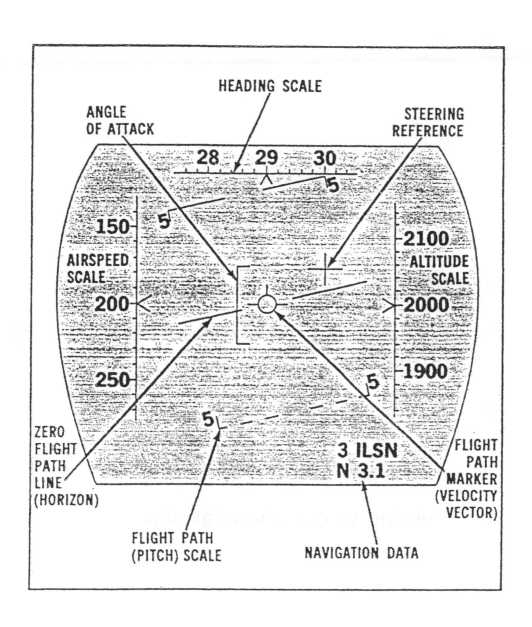

WARNING SYSTEMS

Three basic objectives:

1- Alert

2- Report

3- Guide

WARNING SYSTEMS CLASSIFICATIONS

Four Functional Classes:

1- Performance or departure from safe flight profiles

2- Aircraft configuration

3- Status of aircraft systems

4- Related to communications

WARNING SYSTEMS CLASSIFICATIONS

Four Urgency Conditions:

Emergency Condition

- Warning - Red - Immediate action

Abnormal Condition

- Caution - Amber - Pending emergency

Advise

- Some crew action may become necessary

Alerts

- Provide information

DAY 20   Controls

PO: 1.   Describe the various functions of the controls in the cockpit and give examples of each.

1.   How do aircraft controls fit into the SHEL model for human factors interface?

(HF p. 261)

      -      Liveware ---> Hardware

2.   What are the functions a control may fill?   (HF p. 261)

      a.   Transmit discrete information

             -      Transponder code

      b.   On/Off command

             -      No Smoking lamp

      c.   Transmission of continuous information

             -      Cabin temperature selector

      d.   Control a display directly

      e.   Give a signal to a system

3.   How do controls differ in the forces required to actuate them?   (HF p. 261)

      a.   Electrical   = Lower

      b.   Hydraulic   = Medium

      c.   Mechanical   = Higher

PO: 2.   Explain each of the five design principles for cockpit controls.

4.   What are the five principles of design for cockpit controls?   (HF p. 261-262)

a.   Control-display distance ratio:

-      Distance of control movement relative to the display movement

-      Sensitivity - often a critical design factor effecting operator performance

b.   Direction of movement:

-      Relative to display movement

-      Clockwise to increase

-      Vertical scales -   same as side closest

c.   Resistance:

-      Affects "feel" and smoothness of control movement

-      Four kinds of resistance:

       1-   Elastic

       2-   Friction (static and sliding)

       3-   Viscous dampening

       4-   Inertia

d.   Coding:

-      Improve identification

-      Reduce errors and time for selection

-      By means of: -   Shape
                      -      Size
                      -      Color
                      -      Labeling
                      -      Location

e.   Inadvertent actuation protection:

-      Resistance will help

-      Gating

- Locking devise

- Lever-locked

- Guarded

- Recessed

- Placed out of the way

- Interconnected

- Advisory or warning attachments

PO: 3.   Explain the applications of keyboards in the cockpit.

5.   What are the developments in cockpit technology which bring the use of keyboards as controls to the forefront?   (HF p. 262)

a.   Still the predominant means to instruct machines

b.   Required to give instructions to computerized systems

- Inertia and area navigation systems

- Flight management and data link systems

PO: 4.   *Describe the differences between the QWERTY and DSK or Dvorak keyboard layout.*

6.   How would you describe the differences between the QWERTY and DSK or Dvorak keyboard layout?   (HF pp. 262-263)

a.   QWERTY - Developed in 1878

- Ergonomically poor

- Finger load varies from 1% up to 22%

- Nearly 60 % of work done by the left hand

b.   DSK or Dvorak

- Designed for speed and efficiency of operation

- All 5 vowels and most common consonants in the center row

- Finger loading varies from 8% to 18%

- balance of the work shifted to the right hand

- Not yet widely accepted

PO: 5.   *Explain the design considerations for flight deck applications of keyboards.*

7.   What are the flight deck applications of keyboards which must be considered in the design?   (HF p. 264)

     a.   Accuracy better than speed

     b.   Use in turbulence

     c.   Less than optimal location

     d.   Utilize only one hand

     e.   Confined space

     f.   Numerics should be placed on a separate block

         -      123 on top is faster and more accurate than the 789 on top row

     g.   Key size must be optimized with space and fences between keys

8.   What may be the role of the keyboard in the cockpit of the future?   (HF p. 264)

     -      Pilot's Desk Flight Station is a change from the

         Pilot's Flight Station to the Computer Operator's Station

PO: 6.   Describe the advantages and disadvantages of using automation in the cockpit.

10.   What are the three broad objectives (advantages) cited to justify the introduction of cockpit automation?   (HF pp. 265-266)

     a.   Task demands exceeds human capability for safety and reliability

         -      Aerodynamic characteristics make manual flying impractical

- Stability augmentation systems (SAS)

  - Auto-land

  - Altitude & heading hold

  - Altitude capture

  - Warning systems

b. Economics

  - FMS permit more efficiency

    - Saves time and fuel

    - 3 - 5% depending on trip length

c. Reduce pilot workload

11. What has been one of the more notable human factors problems arising from the development of automatic systems? (HF p. 265, 267-268)

a. Automatic complacency

  - Reliance on the auto-throttle and landing too fast

  - Korean Airlines 007 shoot down

b. Their normal reliability provides the foundation for overconfidence

c. Monotony and boredom which result from under stimulation associated with automation

d. Human requires:

  - Responsibility & Challenge

  - Accomplishment

  - Satisfaction

  - Motivation & Performance

12. What are some of the advanced concepts for the modern cockpit controls of today's, and future air transport aircraft? (HF p. 268)

    a. Touch sensitive CRT

    b. Voice command

    c. Fly-by-wire techniques

    d. Side stick controller

CONTROLS

Function:

- Transmit discrete information

    - Transponder code

- On/Off command

    - No Smoking lamp

- Transmission of continuous information

    - Cabin temperature selector

- Control a display directly

- Give a signal to a system

Force for Actuation:

- Electrical   = Lower

- Hydraulic   = Medium

- Mechanical   = Higher

PRIMARY DESIGN PRINCIPLES

1- Control-display distance ratio:

- Distance of control movement relative to the display movement

- Sensitivity

2- Direction of movement:

- Relative to display movement

- Clockwise to increase

- Vertical scales - same as side closest

3- Resistance:

- Affects "feel" and smoothness of control movement

    - Elastic

    - Friction (static and sliding)

    - Viscous dampening

    - Inertia

4- Coding:

- Improve identification

- Reduce errors and time for selection

- By means of: - Shape
                  - Size
                  - Color
                  - Labeling
                  - Location

5- Inadvertent actuation protection:

- Resistance will help

- Gating

- Locking devise

- Lever-locked

- Guarded

- Recessed

- Placed out of the way

- Interconnected

- Advisory or warning attachments

KEYBOARDS

* Still the predominant means to instruct machines

* Required to give instructions to computerized systems

    - INS or RNAV

    - FMS

    - Data link

* Keyboard layout:

    - QWERTY - Developed in 1878

        - Ergonomically poor

    - DSK or Dvorak

        - Designed for speed and efficiency of operation

        - Not yet widely accepted

*   Flight deck application considerations

-    Accuracy better than speed

-    Use in turbulence

-    Less than optimal location

-    Utilize only one hand

-    Confined space

-    Future possibilities

-    Pilot's Desk Flight Station

Pilot's Flight Station

Computer Operator's Station

AUTOMATION

Advantages & Disadvantages

Advantages:

* Task demands exceeds human capability for safety and reliability

    - Aerodynamic characteristics make manual flying impractical

        - Stability augmentation systems (SAS)

        - Auto-land

        - Altitude & heading hold

        - Altitude capture

        - Warning systems

* Economics

    - FMS permit more efficiency

        - Saves time and fuel

        - 3 - 5% depending on trip length

* Reduce pilot workload

AUTOMATION

Advantages & Disadvantages

Disadvantages:

* Automatic complacency

    - Reliability = > Overconfidence

* Under-stimulation

    - Brings monotony and boredom

* Human requires:

Responsibility & Challenge

Accomplishment

Satisfaction

Motivation & Performance

ADVANCED CONCEPTS

- Touch sensitive CRT

- Voice command

- Fly-by-wire techniques

  - Side stick controller

The HUMAN Operator and AUTOMATION

a.   The Human operator must be:

    1.   Involved

    2.   Informed

    3.   Able to monitor the automated system

b.   The automated system must be:

    1.   Predictable

    2.   Able to monitor the human operator

c.   Each element of the system must know the other's intentions

HUMAN FACTORS IN AVIATION SAFETY
(SF 320)
LECTURE NOTES

DAY 21   Space and Design

PO: 1.   *Describe the aspects of human factors to apply in the design of space in transport aircraft.*

1.   What is the primary task of the human factors specialist in designing working and living space in transport aircraft?   (HF p. 269)

    a.   Matching the characteristics of the human with the working areas

    b.   Characteristics of body parts to consider include:

        -     Size

        -     Shape

        -     Movements

2.   What aspects must be considered in this optimization?   (HF p. 269)

    a.   Performance

    b.   Comfort

    c.   Safety

3.   What areas need to be considered for space and layout design?   (HF p. 269)

    a.   Flight deck

    b.   Cabin

    c.   Maintenance areas

    d.   Cargo compartments

4.   What knowledge must be applied when designing equipment for use on aircraft?
                    (HF p. 269-270)

a.   The user's ability to access or use the equipment considering height, girth, reach, and strength

b.   Equipment should be designed to match the characteristics of people rather than the reverse

PO: 2. Define the terms anthropometry and biomechanics and cite applications.

5.   What are the two disciplines involved in matching hardware, software and the environment to the characteristics of the human?

(HF p. 270)

a.   Anthropometry

b.   Biomechanics

6.   What is anthropometry concerned with?   (HF p. 270)

a.   Human dimensions   -   Size of different limbs
                             -    Weight
                             -    Stature
                             -    Seated eye height
                             -    Reach

b.   Anthropometer   -   Developed to facilitate measurement

c.   Calculate   -   Optimum height of a work surface
                             -    Location of controls
                             -    Height and depth of stowage areas
                             -    Minimum knee room between seats
                             -    Width of seats
                             -    Length of arm rests
                             -    Height of headrest

d.   Dynamic anthropometry   -   Involves:
                                      -    body movements
                                      -    reach requirements
                                      -    working space

7.   What does is the study of biomechanics concerned with in human factors? (HF p. 270)

a.   The forces which the body parts can apply

b.   Direction which movements have to be made

PO: 3.   *Explain the various aspects of human dimensions as they relate to aircraft design considerations.*

8.   How is information collected about human dimensions?   (HF p. 271)

   a.   Skill in measurement

   b.   Use of the anthropometer which measures:

          -     stature
          -     arm reach
          -     sitting eye height
          -     various body dimensions

9.   What care must be taken by the researcher when establishing a standard for human dimensions in the design of equipment?   (HF p. 271)

   -     Obtaining a large enough representative sample of people who will use the equipment

10.   What else must be considered when designing for human dimensions?   (HF p. 271)

   a.   You may have to place a size limitation in the selection   screening

   b.   Human physical dimensions have slowly increased

          -     Males   = 1.3 mm/year

          -     Females   = 0.9 mm/year

   c.   May have to design equipment for 30 year use

          = 30 mm (1 1/8") change in the average

11.   What other differences must the designer of equipment for worldwide use consider?
                                                                              (HF p. 271)

   a.   Ethnic differences

          -     Negro   =   Longer legs compared to Caucasians

          -     Asians   =   Smaller overall with longer trunk and shorter legs compared to Europeans

   b.   Sex differences

Refer to Fig 12.1 ----- > (HF p. 272)

c.   Must ensure adequate foot load can be applied while sitting at the design eye
position of seat adjustment

-      Not to exceed 150 lbs required at minimum control speed with engine
failure (US airworthiness standards)

12.   What is the U.S. transport category aircraft design requirement for accommodating
crew size ? (HF p. 272)

-      Range =   1.57 m - 1.90 m
(5'-2") - (6'-3")

PO: 4.   *Describe how the study of statistics works to aid the human factors considered
in aircraft space design.*

13.   What statistical concepts have value in arranging the results of anthropometric studies
into meaningful design guidance?

(HF p. 273)

-      Most human body dimensions approximate to the shape of a familiar
mathematical distribution

Gaussian distribution or typical bell curve shape

14.   What are the two main indices required for the description of the design population?

Refer to Fig 12.2 ----- > (HF p. 273)

a.   Mean

b.   Standard deviation

15.   What are the general human factors guidelines used for space design?   (HF p. 274)

a.   Large people determine clearance

b.   Small people determine limits for reach

c.   Must provide adjustment range

16. Where does the statistical concept of percentile fit into the design for space in aircraft?
(HF p. 274)

   a.   The percentile is a means of expressing the range of sizes to be accommodated

   b.   Determines those who will be left out of the size capability

   c.   Accommodating a very large percentage of the population would be very expensive

   -   Pilot seat adjustment example:

Population Included		Adjustment Range Required
90%	=	11 cm (4.3")
98%	=	15 cm (5.9")
100%	=	26 cm (10.2")

   d.   A fundamental design decision is the decision on how many disadvantaged minority will be excluded

PO: 5.   Describe some of the constraints imposed on the proper human factors design of cockpit and cabin space.

17. What has been the difficulty in the past with the design of the cockpit space?
(HF pp. 274-275)

   a.   Designers were lacking in an overall view of the flight crew

   b.   Designers were lacking formal human factors training

   c.   There is a strong case for all design engineers to undergo some basic education in human factors

   d.   It's much better to design out the human factors problems in the conceptual stage rather than to find them in the development stage and then have to fix them

Design for Human Factors

Conceptual Stage   VS   Developmental Process

18. What constraints often limit the extent the designer can optimize hardware?

(HF pp. 275-276)

    a.   Aerodynamic characteristics

        -    Cross section of the fuselage and the shape of the nose

            -    Concorde narrow   = 148 cm (57")

            -    Boeing 747 wide   = 191 cm (75")

    b.   Commercial pressures

        -    Cabin space is sold

        -    Flight deck and galley space is not sold

19. How do visibility requirements influence design for space?   (HF p. 276)

    a.   Downward visibility requirement influences windscreen design and location.

    b.   Design eye location determines

        Seat location

            Seat location determines

                -    Rudder & control yoke location

                -    Instrument and control panel location

20. What considerations must be made for designing the space between the pilots?

(HF p. 276)

    a.   Closer =   Better cross-monitoring
                    -    Loss of pedestal panel space
                    -    Restricted inboard access

    b.   Farther =   Better outside lateral visibility
                    -    More use of pedestal panel space
                    -    More inboard access

    c.   Large jets   = 105 cm   (41.3")

    d.   Concorde   = 89 cm   (35")

PO: 6.   *Describe the historical development and principles of design for layout of cockpit display panels.*

21.   What are the types of cockpit display panels that need to be considered for human factors optimization?   (HF p. 277)

     a.   Flight instrument

     b.   System quantitative

     c.   System displays and controls

     d.   Flight guidance control

     e.   Radio and interphone

22.   What are the cockpit geometry concerns for panel displays that are associated with viewing distance?

         Refer to Fig 12.3 ----->   (HF pp. 277-278)

     a.   Before size is determined

         -      Location

         -      Who will use

     b.   Main instrument panel is about 71 - 78 cm (27.9 - 30.7")

     c.   Overhead panel as close as 20 cm (7.9")

     d.   Lateral systems panel as far as 2 m (79")

23.   What were the historical developments in improving the design of cockpit instrument panels?   (HF p. 277)

     a.   James Doolittle (1930s) first serious consideration for display organization

     b.   Standard Blind-Flying panel by the RAF in 1937

         Refer to Fig 12.4 ----->   (HF p. 261)

         -      Basic T

-       For basic scanning of the four basic parameters:

    1.   Attitude

    2.   Speed

    3.   Altitude

    4.   Heading

24.   What considerations should be given to the design of system quantitative information?
(HF p. 278)

a.   Normal position of pointer = 9 or 12 o'clock

b.   Pointers with extended tails are better than short ones

c.   The panel should be organized as a symmetric pattern of the whole, so deviations will stand out

25.   What systems often use the mounting of displays and controls in a schematic or synoptic form?
(HF p. 278)

a.   Fuel

b.   Electrical

c.   Pneumatic

d.   Hydraulic

f.   Current technology permits touch sensitive CRT operation

26.   What are some of the basics for design of the flight guidance control panels?
(HF p. 278)

a.   Generally mounted on the glare shield

b.   Digital readouts

c.   Reach for operational controls

d.   Spacing and coding are important for blind or peripheral vision

27.   What other control and display panels in the cockpit or elsewhere require optimization?
(HF p. 278)

a.   Radio and interphone

b.   Circuit breaker

c.   Galley and door operation

28.   What are the two methods for determining the direction for toggle switch movement?

Refer to Fig 12.5 ----->   (HF p. 279)

a.   Forward-on

b.   Sweep-on

29.   How does the sight and reach principle apply to cockpit display and control panels?

(HF p. 280)

a.   There may be a different optimum location for

-       displays (viewing) and controls (reaching)

b.   Displays and controls in a single module

-       Nice for engineering and logistical convenience

-       Frequently not optimum for human factors

Refer to Fig 12.6 ----->   (HF p. 280)

c.   Flight display located on the instrument panel

-       with controls located on the glare shield

PO: 7.   *Describe the way the crew complement on an aircraft influences the layout of the flight deck.*

30.   How does the crew complement influence the layout and design of the flight deck?

(HF pp. 281-282)

a.   Two or three crew members is fundamentally important in the basic design process

b.   Crew Placement (Third crew member)

-       Must be established at the drawing board stage

- Forward facing - Seems better for the transition from crew of three

- Side facing - the rule in all wide bodied jets

c. Extensive overhead panel

- Necessary for the transition to the crew of two

- Forward section - more frequently used items

- Rear (less convenient) section - less frequent use

d. Whatever the crew complement the location of the crew member and the panels must be established at the drawing board stage of development

PO: 8. *Describe design considerations for the crew seats on transport aircraft.*

31. What are the requirements which drive design considerations for flight deck crew seats? (HF p. 282-283)

a. Few flight deck components have more complaints

b. Limited leg movement for long hours = aggravates circulation

c. Conducive to unhealthy posture

- Frequent back pain or discomfort

- Lumbar disc disorders

d. 30 years of exposure is possible

e. Development of seats for long range use must be tested in the long range environment

Refer to Fig 12.7 -----> (HF p. 284)

f. Total adjustment range = approx 7"

32. What considerations should be given to cabin crew seat and facility design? (HF p. 284)

a. Seats designed and positioned for prevention of injury during emergency landings or from flying objects

b.   Prevention of back injuries from:

-    inadequate work surfaces

-    badly designed stowage facilities

-    unsecured doors on stowage areas

c.   Adequate storage for cabin crew personal luggage

SPACE AND DESIGN

Matching components

    Human < ----> Working Areas

    -    To optimize

        -    Performance
        -    Comfort
        -    Safety

Areas to consider space and layout

    -    Flight deck

    -    Cabin

    -    Maintenance areas

    -    Cargo compartments

ANTHROPOMETRY

- Human dimensions: - Size of different limbs

    - Weight

    - Stature

    - Seated eye height

    - Reach

- Anthropometer

    - Developed to facilitate measurement

- Calculate:
    - Optimum height of a work surface

    - Location of controls

    - Height and depth of stowage areas

    - Minimum knee room between seats

    - Width of seats

    - Length of arm rests

    - Height of headrest

BIOMACHANICS

- The forces which the body parts can apply

- Direction which movements have to be made

- Human physical dimensions slowly change

    - Males   = 1.3 mm/year

    - Females   = 0.9 mm/year

- Ethnic differences

    - Negro   =   Longer legs compared to caucasians

    - Asians   =   Smaller overall with longer trunk and shorter legs compared to Europeans

Transport category crew size range =

1.57 m - 1.90 m
(5'-2") - (6'-3")

- Human Factors Statistics

    - Required indices

        - Mean

        - Standard deviation

- General Human Factors guidelines

    - Large people determine clearance

    - Small people determine limits for reach

    - Must provide adjustment range

    - Percentile - a means of expressing the range of sizes to be accommodated

COCKPIT DESIGN

Design for Human Factors

Conceptual Stage   VS   Developmental Process

- Design Constraints

    - Aerodynamic characteristics
    - Commercial pressures

- Visibility requirements influence:

    - Windscreen design

    - Design eye location -->

        Seat location -->

            Rudder & control yoke location

            Instrument & control panel location

- Distance between pilots:

  - Closer = - Better cross-monitoring

    - Loss of pedestal panel space

    - Restricted inboard access

  - Farther = - Better outside lateral visibility

    - More use of pedestal panel space

    - More inboard access

- Large jets = 105 cm (41.3")

- Concorde = 89 cm (35")

- Panel Displays

  - System quantitative

  - System displays and controls

  - Flight guidance control

  - Radio and interphone

  - Before size is determined

    - Location

    - Who will use

  - Main instrument panel is about 71 - 78 cm

    (27.9 - 30.7")

  - Overhead panel as close as 20 cm

    (7.9")

  - Lateral systems panel 2 m

    (6' -7")

Panel Types:

- Flight instrument

    - Standard Blind-Flying

    - Basic T

Crew Complement (Two or Three)

- Fundamentally important in the basic design process

Crew Placement (Third crew member)

- Must be established at the drawing board stage

- Forward facing

- Side facing

HUMAN FACTORS IN AVIATION SAFETY
(SF 320)
LECTURE NOTES

DAY 22   Human Factors in the Cabin

PO: 1.   *Describe the human factors to consider in the design of the cabin environment in terms of considerations given for the passengers and the crew.*

1.   How important are the human factors in the cabin?   (HF p. 286)

    a.   The interface between the cabin attendants and the passengers (human payload) is fundamental

    b.   The human payload = (revenue source < - - - - - > airline)

2.   What areas of the cabin must be considered for human factors   optimization?
                                                               (HF p. 288)

    -     Emergency evacuation

    -     Language barriers

    -     Physical handicaps

    -     Understanding human behavior in emergencies

    -     Rare to find new sources of error or injury

3.   What fundamental difference is there concerning the cabin environment between the cabin crew and the passengers?

        Refer to Fig 13.1 - - - - ->   (HF p. 686)

Cabin Crew	vs	Passengers
Workplace		Rest place
Active		Inactive

4.   What governs the priority of considerations between cabin crew and passengers?
                                                             (HF p. 287)

Comfort & Convenience   ---> Passengers

Safety   ---> Cabin Crew

5.   How does anthropometry and biomechanics fit into cabin human factors design?

                                                        (HF p. 288)

- Body measurements and movements

  - Work-space layout

  - Equipment design

  - Seat design

  - Height of overhead luggage racks and doors

  - Space between seats

  - Door controls

  - Escape facilities

6.   What should be the primary motivation for the flight attendant?   (HF p. 288)

- Satisfaction of working with people

7.   What is one of the great challenges of working with people in the cabin?

                                                          (HF p. 288)

a.   Understanding the individual differences of the people (liveware) there

- Cultural differences

- Psychological differences

  - Tolerance to circadian rhythm disturbances

b.   All must be accommodated for if safety and efficiency of operation are to be maintained

PO: 2.   Describe the major regions of the cabin requiring human factors optimization.

8.   What is a major cabin region demanding human factors attention in matching the hardware to the liveware? (HF p. 289)

- The galley - where the injury risks are greater

   - Heat and fire

   - Electrical appliances

   - Sharp edges

   - Service lift

   - Service trolley

9.   Upon what is efficiency in the galley dependent? (HF p. 289)

   a.   General space and layout

   b.   Number and location

   c.   Control panel design

   d.   Design of individual components

   - Tray stowage

   - Service trolley

10.   To what does delethalization refer? (HF p. 289)

   a.   Process of removing sharp objects or protrusions

   b.   Insuring adequate retention of loose equipment

   c.   FAA requires securing of all:

   - Galley equipment including serving trolleys

   - Crew luggage

   - Not a hazard by shifting in emergency conditions

11. What special human factors considerations must be made when designing emergency equipment? (HF p. 290)

   - Human behavior under conditions of stress

12. What does the inadvertent actuation of an emergency slide cost? (HF p. 290)

   $ 5,000 - $ 15,000

13. What are the problems associated with designing emergency equipment? (HF p. 290)

   a. May need to be used in total darkness

   b. May need to be operated by passengers unfamiliar with it

   c. Environment may be chaotic

   d. FAA tests in 1983-84

      - 1/3 of the passengers could not don their life-vests successfully, even after viewing a demonstration

PO: 3. *Describe factors associated with the use of medical kits on board transport aircraft.*

14. What are the two priority actions a flight attendant should take in the event of a medical emergency in the cabin? (HF p. 291)

   a. Notify the captain

   b. Determine if a doctor or a nurse is on board

15. What are the most troublesome conditions occurring to passengers in flight? (HF p. 291)

   a. Gastrointestinal

   b. Unconsciousness

   c. Shortness of breath

   d. Chest pains

16.   How frequent are in-flight deaths?   (HF p. 291)

   a.   577 during the period 1977 - 1984

   b.   About 82 per year average for the 7 years

PO: 4.   Describe the causes of the fire hazards associated with the cabins of transport aircraft and corrective actions being taken to reduce them.

17.   What are the major hazards in crash survivable accidents?   (HF p. 292)

   a.   The effects of fire account for 30 - 40% of the fatalities in survivable accidents

   b.   Toxic smoke from burning polyurethane foam materials

      -   80% of all transport aircraft fire fatalities are due to toxic gasses rather than thermal injury

   c.   Polyurethane burning --> hydrogen cyanide and carbon monoxide

18.   What requirements are now in force to reduce the toxic gasses in aircraft fires?
                                                                            (HF p.292)

   -   Since 1987 all large aircraft built or operated in the USA or UK are required to install fire-blocking seat cushion covers

19.   What actions must be taken to reduce the loss of life due to cabin fire?   (HF p. 293)

   -   Train cabin crew to fight fires

   -   Develop an aggressive approach to fire

   -   Take effective measures to protect passengers

   -   Captains must plan to land at the nearest suitable runway

   -   Effective communication between the cabin and the flight deck is vital

PO: 5.   Explain the important human factors considerations for cabin seat design.

300

20. Why is the cabin seat so important to a passenger airline?   (HF p. 293)

    a.   It is the airline's interface with the customer

    b.   Passengers spend many hours of confinement there

21. What seat design considerations must be made?   (HF pp. 293-294)

    a.   Adjustment capacity for multipurpose use

    b.   Body weight distribution from seat pan contouring

    c.   Seat height   43 cm (17")

    d.   Seat width 40 cm (16")

    e.   Built in lumbar support

    f.   Armrest   -   height adjustable or 18 - 23 cm (7 - 9")

22. What is seat pitch?   (HF p. 294)

    a.   The spacing between the seats (leg room)

    b.   Normally 71 cm - 86 cm   (28" - 34")

        Refer to Fig 13.2 ----->   (HF p. 295)

23. What G loading requirements do seats have?

        Refer to ----->   (Table 13.1, HF p. 295)

    a.   Recommend = 20-25 g   FAA minimum = 9 g

    b.   Downward = 15-20 g   FAA minimum = 4.5 g

24. What increase in G load force is derived from upper body restraint?   (HF p. 294)

    -    Forward load tolerance can be doubled

25. What requirements do cabin attendant jump seats have? (HF p. 296)

    a. Near emergency doors

    b. Shoulder harness and seat belt restraint

    c. Giving a view cabin area

    d. Located to minimize flying object injury

26. What is the human factors difference between forward and rear facing seats?
(HF p. 296)

    a. Greatly increased survival with reward facing seats

    b. Loads as high as 83 g in reward facing seats

27. What other furnishings in the cabin must also conform to human factors design considerations? (HF p. 297)

    a. Entertainment system

    b. Lighting

    c. Ash trays

    d. Overhead luggage containers

        - 78% failure rate in survivable accidents

    e. Toilet

        - Handicapped provisions

        - Smoke detector and fire extinguisher installation

PO: 6. *Give examples of software items in the cabin which require human factors optimization.*

28. What are some examples of the items of cabin software requiring human factors optimization? (HF p. 298)

    a. Instructions

    b. Checklists

c.   Passenger briefing cards

d.   Cabin placards

PO: 7.   Describe the broad categories of cabin duties and explain the human factors issues to be considered in them.

29.   Into what two broad categories are cabin staff duties divided?   (HF p. 298)

   a.   Operational Tasks:

      -       Determined by regulation

      -       Administrative paperwork

      -       Checking and preparation of flight safety equipment

   b.   Service Duties:

      -       Established by the marketing department

      -       Meal provisions

      -       Interaction with passengers

30.   Who in the cabin seems to suffer more injuries from in-flight turbulence?
                                                                  (HF p. 298)

   a.   From 1979 - 1983   -   65% were flight attendants

   b.   Usually 30 times more passengers

31.   What is the reason they suffer more of the in-flight turbulence injuries?   (HF p. 298)

   a.   They feel obligated to continue service as long as possible

   b.   Sometimes leave serving trolleys and loose equipment un-stowed when they should be put away

      -       Typical service trolley weight:

         -       Empty   = 35 kg (77#)

         -       Full   = 90 kg (198#)

- Techniques are required for rapid security

32. How important is the neatness and general appearance of the cabin staff in the Liveware-Liveware interface? (HF p. 299)

    a. They represent the airline's shop window to the public

    b. They have a profound impact on the company's image

    c. Staff/Passenger ratio = 1:20 - 40 (QE II = 1:2)

PO: 8. Identify each of the cabin environmental areas and describe how human factors is applied to each.

33. What are the factors of concern in improving the interface between the Liveware and Environment components of the cabin? (HF p. 301)

    a. Noise

    b. Temperature, humidity and pressure

    c. Ozone

    d. Smoking

    e. Circadian and time zone effects

34. What are the sources of noise in fixed wing transport aircraft? (HF p. 301)

    a. Boundary layer turbulence

        - Increases with speed

        - Greater at the front of wide bodied jets

    b. Air conditioning and pressurization systems

    c. Engines

    d. Hydraulic systems

35.   What are the noise sources in helicopters?   (HF p. 301)

    a.   Aerodynamic from the rotor

    b.   Gearbox and drive mechanism

36.   What are the effects of noise on those subject to the environment of transport aircraft?
                                                                       (HF p. 302)
    a.   Impairment of hearing

       -    Crew hearing loss is more than normal

    b.   Masking of alerting signals

    c.   Annoying for most people

    d.   Increase in workload and fatigue

    f.   Passengers are disturbed by changes in noise

37.   What human factors considerations arise from temperature and humidity aspects of the cockpit/cabin environment?   (HF p. 303)

    a.   Temperature control most significant area of conflict between passenger and crew needs.

    b.   Air at altitude has very low humidity

          Refer to Fig 13.3 ----->   (HF p. 304)

       -    The amount of passengers effects humidity

          -    More passengers = more humidity

       -    Increase fluid intake to reduce dehydration

       -    Coffee and alcohol should be avoided (diuretics)

38.   What are the human factors of pressurization to consider in the environment of the cabin?   (HF p. 305)

    a.   Normal pressure is 5,000 - 8,000 ft

    b.   Loss of pressurization

- Time of useful consciousness (TUC) depends on:

  - Aircraft altitude

  - Rate of decompression

  - Activity of occupant

- Flight and cabin crew must be on supplemental oxygen before TUC expires

- Time of safe unconsciousness (TSU):

  - How long a person can remain unconscious without risk of brain damage from lack of oxygen

  - About 2 minutes

39. What are the human factors considerations for ozone in the cabin environment?
(HF p. 306

   a. Created by the action of the sun on oxygen

   b. It is a toxic gas which damages the lungs through emphysema

   c. Concentration increases above the tropopause (above 34,000')

   d. Concentration increases with increased latitudes

   e. Damage appears to come more from concentration rather than period of exposure

   f. Breathing through a damp cloth might help with the chest discomfort and coughing associated with inhalation of ozone

40. What human factors issues on aircraft come from smoking?   (HF pp. 307-308

   a. One of the main causes of emotional friction in the cabin

   b. Attitudes have become less tolerant

   c. Eye nose and throat irritation are worse with low humidity

   d. Some air conditioning systems recirculate a substantial part of the cabin air

e.   Fire hazard, burn damage, and damage from tobacco tar are also issues to consider

41.   What considerations in the cabin must allow for the circadian and time zone effects?

(HF p. 311)

a.   Cabin crew may have difficulty sleeping

b.   Scheduling of meals and in-flight movies must take into account not only local time but the body phase of the majority of the passengers

HUMAN FACTORS IN THE CABIN

LIVEWARE <--> ENVIRONMENT

Revenue source <-----> Airline

CABIN PRIORITY

<u>Cabin Crew</u>   vs   <u>Passengers</u>

Workplace   Rest place
Active   Inactive

Comfort & Convenience ---> Passengers

Safety   ---> Cabin Crew

EMERGENCY EQUIPMENT
DESIGN CONSIDERATIONS

a. May need to be used in total darkness

b. Passengers bay be unfamiliar with operation

c. Environment may be chaotic

TROUBLESOME CONDITIONS OCCURRING
TO
PASSENGERS

a. Gastrointestinal

b. Unconsciousness

c. Shortness of breath

d. Chest pains

TO REDUCE LOSS OF LIFE FROM CABIN FIRES

-   Train cabin crew to fight fires

-   Develop an aggressive approach to fire

-   Take effective measures to protect passengers

-   Plan to land at the nearest suitable runway

-   Effective cockpit - cabin communications

PASSENGER SEAT DESIGN

a. Adjustment capacity for multipurpose use

b. Body weight distribution from seat pan contouring

c. Seat height 43 cm (17")

d. Seat width   40 cm (16")

e. Built in lumbar support

f. Armrest   -   height adjustable or 18 - 23 cm
                                                    (7 - 9")

CABIN ATTENDANT JUMP SEATS

a. Near emergency doors

b. Shoulder harness and seat belt restraint

c. Giving a view cabin area

d. Located to minimize flying object injury

CABIN STAFF DUTIES

a. Operational Tasks:

b. Service Duties:

ENVIRONMENT IN THE CABIN

- Noise

- Temperature, humidity and pressure

- Ozone

- Smoking

- Circadian and time zone effects

SOURCES OF NOISE

a. Boundary layer turbulence

b. Air conditioning and pressurization systems

c. Engines

d. Hydraulic systems

EFFECTS OF NOISE

a. Impairment of hearing

b. Masking of alerting signals

c. Annoying for most people

d. Increase in workload and fatigue

f. Passengers are disturbed by changes in noise

HUMAN FACTORS IN AVIATION SAFETY
(SF 320)
LECTURE NOTES

DAY 23   Interface Between People

PO: 1.   Identify some of the considerations that must be given to communications in the cockpit and cabin.

1.   How important is communication in the activities of the cabin?   (HF p. 311)

  -      It is a major part

2.   What are some of the barriers to communications between the flight deck and the cabin?   (HF pp. 311-312)

  a.   Cockpit door

  b.   Requirement for pilots to be strapped in their seats

  c.   Flight deck "sterile" times

    -      Take-off, climb, descent and landing

  d.   Frequently a different leader and crew on each flight

3.   What type of problems can decrease the effectiveness of the public address (PA) system?   (HF pp. 312-313)

  a.   Difficult to design properly

  b.   Difficult to maintain adequately

  c.   Vast variation in ability of crew to use it effectively

4.   What unique position does any crew member seem to be in when they walk down the aisle?   (HF p. 313)

  a.   All eyes of the passenger are upon them

  b.   Body language cues can communicate concern or lack thereof

5.   What special physical handicaps to communication in the cabin must be accounted for?
                                                                                (HF pp. 313-314)

        a.   Hearing deficiency   -   Suffered by 1/6th of the population

        b.   Visual impairment

        c.   Regulations:

                -       Permit guide dogs and wheel chairs on the aircraft

                -       Prohibit unjust discrimination against the handicapped

PO: 2.   *Describe the requirements for dealing with intoxicated passengers.*

6.   What are the regulatory requirements with regard to intoxicated passengers?
                                                                                (HF p. 314)

        a.   FAA prohibits an airline from allowing on board any person who appears to be
                intoxicated

        b.   Airlines are reluctant to ban marginal cases

        c.   No breath test is required for boarding

PO: 3.   *Describe the considerations associated with a passenger's fear of flying, and
                cite measures which can help reduce this human response.*

7.   How does fear of flying or flying phobia fit into the cabin Liveware-Liveware
        interface?
                                                                                (HF pp. 315-316)
                Refer to Fig 13.6 ----->   (HF p. 316)

        a.   Almost a third of the population have some anxiety or fear

        b.   Individuals differ on what phase is greatest

                -       Take-off for most

                -       Turbulence and maneuvers also

8.   How can you best treat the fear of flying?   (HF pp. 317-318)

                        Refer to Appendix 1.26 pp. 349-350

a. Sometimes other phobias are simultaneous and requires a multi-dimensional approach

   -   Fear of heights (acrophobia)

   -   Fear of closed (tight) places (claustrophobia)

b. Education is usually helpful

   -   There is a marked ignorance of elementary theory of flight and meteorology

c. Enlightened use of the PA system can help

9. What benefits can come from recognizing and accounting for fear of flying?
                                                                    (HF p. 318)

   a. More effective passenger interface

   b. More revenue passengers

PO: 4.   Describe the aspects of passenger behavior problems in the cabin, and give practical measures to deal with them.

10. What passenger behavior in the cabin can have serious safety implications?
                                                                    (HF p. 319)

   a. Abuse and assault by passengers

   b. FAA specifically forbids interference with the crew

      -   Provides legal basis for action

   c. Most airlines are reluctant to proceed with prosecution

11. What are some practical measures that can be taken against passenger abuse/assault of the flight crew?   (HF p. 319)

Refer to Appendix 1.27 p. 350

PO: 5.   Describe the nature of passenger behavior in an emergency, and what crew members can do to enhance successful emergency response.

12. What are the characteristics of flying that are unique with regard to behavior in an emergency?   (HF p. 320)

a.   Passengers feel powerless to control events that impact on their destiny

b.   In flight:

-    Enclosed in a confined space

-    Rather densely packed

-    No escape routes in flight

c.   On the ground:

-    May require somewhat athletic use of escape slides

-    Visibility can rapidly drop to near zero

-    May be in a structurally damaged cabin

-    Widespread disorientation may prevail

13.   What important Liveware-Liveware principles are highlighted from the feelings flying passengers have about their situation?   (HF p. 320)

a.   Passengers look to the crew for leadership, guidance and instructions to insure escape, survival and rescue.

b.   A group becomes more dependent on a leader in stress situations.

c.   The PA systems or megaphones may be the only way to provide guidance to passengers.

PO: 6.   *Identify the factors highlighted by the statistics related to survivability.*

14.   What survivability statistics come from analysis of aircraft accidents?   (HF p. 321)

-    Seating close to an emergency exit will enhance chances

-    Very young and very old are more vulnerable

-    Females are not as likely to survive

15.   What areas of concern remain with regard to important factors in survivability?
                                                                        (HF p. 322)

- Passenger safety information facilities

- Crashworthiness

- Flight attendant duty time

- Excessive cabin baggage

- Galley equipment

- Training standards

PO: 7.   *Describe the three classes of people who might be encountered, and what the primary tasks of the crew might be in a violent hijacking situation.*

16.   What are the three classes of people who might be encountered in a violent hijacking?
(HF pp. 322-324)

     a.   Terrorist   -   most ruthless, skilled and dangerous

     b.   Criminal   -   martyrdom is not a common characteristic

     c.   Mentally unbalanced -   unpredictable and usually acting alone

17.   What is the primary task of the crew during a hijacking?   (HF p. 325)

     a.   Lower tension

     b.   Be unprovocative

     c.   Establish credible, reasonable and non-controversial communication with the hijacker

INTERFACE BETWEEN PEOPLE
IN THE CABIN

LIVEWARE <------> LIVEWARE

Communications

Special Physical Handicaps

Intoxicated Passengers

Fear of Flying

Passenger Behavior Problems

Emergency Situation Behavior

Survivability

CHARACTERISTIC EMERGENCY BEHAVIOR
IN
FLYING

- Feeling of being powerless to control events that impact on their destiny

- In flight:

    - Enclosed in a confined space

    - Rather densely packed

    - No escape routes in flight

- On the ground:

    - May require somewhat athletic use of escape slides

    - Visibility can rapidly drop to near zero

    - May be in a structurally damaged cabin

    - Widespread disorientation may prevail

LIVEWARE - LIVEWARE PRINCIPLES
FOR
AIRCRAFT EMERGENCIES

- Passengers look to the crew for leadership, guidance and instructions to insure escape, survival and rescue.

- A group becomes more dependent on a leader in stress situations.

- The PA systems or megaphones may be the only way to provide guidance to passengers.

THREE CLASSES OF HIJACKER

1- Terrorist   -   most ruthless, skilled and
                             dangerous

2- Criminal   -   martyrdom is not a common
                             characteristic

3- Mentally   -   unpredictable and usually
   unbalanced   acting alone

PRIMARY TASK OF THE CREW
DURING A HIJACKING

a. Lower tension

b. Be unprovocative

c. Establish credible, reasonable and non-controversial communication with the hijacker

HUMAN FACTORS IN AVIATION SAFETY
(SF 320)
LECTURE NOTES

DAY 24　　　Education and Applications

PO: 1.　　*Describe why there seems to be a barrier between the knowledge obtained in human factors research and the practical application of human factors principles.*

1.　　What are the reasons for the communication breakdown between the academia of human factors and its application?　　　　　　　　　　(HF pp. 326-327)

Human Factors Research	--------------------->	Human Factors Applications
Acidemia		Industry
Knowledge	---------------------->	Practical Problems

　　a.　　Lack of human factors technology knowledge

　　　　-　　at the management level

　　　　-　　In the design office

　　b.　　Lack of funding and support for human factors activity

　　c.　　Lack of proper human factors effort during system development

　　d.　　Academia doesn't usually translate research findings into operationally usable language or techniques

　　f.　　It is difficult to evaluate the cost benefits of an investment to better human factors

PO: 2.　　*Describe the basic educational preparation for human factors work and the specific qualifications for each of the four levels of human factors expertise.*

2.　　What background seems to be the most common for human factors specialists?
　　　　　　　　　　　　　　　　　　　　　　　　　　　　(HF p. 327)

　　a.　　Engineering, industrial or experimental psychology

　　b.　　A variety of psychology-based academic courses

3.   What are the four levels of qualifications for working human factors?

Refer to Fig 14.1 ----- >   (HF p. 328)

Level 4   University higher degree ---- > Consultant
                                +
                        Industry research
                                    -- * --

-        Only called upon when needed

-        Able to make analyses and give advise without internal pressures and biases

-        Routine contact with industry and research developments

^^^

(HF p. 328)

Level 3   University degree   ----> In-house Specialist

-        Available on demand to solve human factors problems

-        Needed expertise for calling on an external consultant

-        Able to carry out research studies

-        Must acquire a full understanding of the work situation

-        Establish a credibility with the operations department

-        To assist in establishing some degree of human factors education at the
         supervisory level

^^

(HF p. 329)

Level 2   Two Week   ----> Supervisors
               Compressed course

-        To enable them to carry out their own tasks effectively

-        Able to recognize human factors problems and solve straight forward ones

-        Understand the kind of support and assistance available

-        Provided by a university, technical institute or organization specializing in
         human factors

Level 1   Awareness course   ----> All staff

- Staff feedback on operational deficiencies helps identify where corrections are needed

- All basic technical training courses should include basic human factors familiarization

- A special human factors familiarization course should be used as an interim measure and be considered for periodic refresher training

^^

4.   What are the criteria used in the development of the KLM Royal Dutch Airlines human factors awareness course?   (HF p. 331)

    a.   Awareness level coverage for the full range of human factors applicable in air transport operations

    b.   Maximum staff coverage at minimum cost per person

    c.   Hardware mobility with minimal logistical cost

    d.   High credibility from Human Factors design quality

    e.   Format and content - suitable for use by all staff

Course design:

- 15 Separate 40 minute self-contained audio-visual units

- Uses air transport examples to illustrate principles

PO: 3.   *Describe some of the specific human factors training which airlines can develop and what they might be designed to accomplish.*

5.   Why might an airline develop a specific human factors training program?
      (HF pp. 331-332)

    a.   Designed to meet a particular need within an operational environment

    b.   May not be suitable for adoption elsewhere

6. What are some of the recent team training and testing concepts in the air transport industry? (HF p. 332)

    a. Line-Oriented Flight Training (LOFT)

    b. Cockpit/Crew Resource Management (CRM)

7. How would you describe Line-Oriented Flight Training (LOFT)? (HF p. 332)

    a. Training and testing of flight crews as a team rather than as individuals

    b. Skills must be practiced as a group activity

    c. Personality and situational factors interact

    d. Particularly oriented towards the L-L interface

    e. Video-taping of the training session for self assessment

PO: 4. *Explain the concept of Crew Resource Management (CRM), describe what it involves, cite what it is designed to accomplish, and specify what it will not do.*

8. What is involved in CRM? (HF p. 332)

    a. The management and utilization of all resources available on the aircraft

        - People

        - Equipment

        - Information

    b. A special case of SHEL system management

9. What are the two general assumptions which are at the origin of CRM? (HF p. 332)

    a. Accidents occur because the resources available are not adequately managed

    b. Resources are available to prevent accidents if they are optimally utilized

10. What are the elements of CRM? (HF p. 333)

a. Leadership

b. Communication

c. Task distribution

d. Setting priorities

e. Monitoring

- Information sources

- Individual performance

f. Extraction and utilization of data

11. What can CRM do? (HF p. 333)

a. Very effective in teaching management skills

b. Many attitudes can be modified

12. What will CRM not do? (HF p. 333)

a. Change personality traits

b. Change the effect on performance of:

- Domestic stress

- Fatigue

- Circadian rhythm disturbance

- Drugs or medication

- Errors from information processing

- Visual illusions

- Design induced errors

PO: 5. *Describe how applications of human factors should be made in the six stages of design and production.*

13. In what areas must human factors applications be applied to receive optimum effectiveness in the air transport industry? (HF p. 334-339)

    a.   Aircraft hardware

    b.   Training

    c.   Operating procedures and documentation

    d.   Staff/Management relations

    e.   Accident and incident investigation

    f.   The regulatory authority

    g.   Marketing

14. Who is responsible for making sure human factors is applied in the design and production of resources within the air transport industry? (HF p. 334)

    a.   Designer

    b.   Manufacturers

    c.   The customer or operator (Airline)

    d.   The state certifying authority

15. What are the six stages of aircraft hardware design where human factors must be applied? (HF p. 335)

    1-   System planning

    2-   Pre-design

    3-   Detail design

    4-   Production

    5-   Test/Evaluation

    6-   Operations

PO: 7. *Describe how human factors expertise is designed into training, operating procedures, staff/management relations, accident investigation, the regulatory authority and marketing.*

16. How does the design of training fit into the human factors picture? (HF p. 335)

  a. Largely an operator responsibility

  b. Certifying authority - approves

  c. Manufacturer - gives initial training

  d. Represents a major cost item

17. Who has responsibility for designing human factors into operating procedures and documentation? (HF p. 336)

  a. Basic operating procedures

    - Designed by the manufacturer

    - Approved by certifying authority

  b. Primary responsibility is the operator or airline under certifying authority supervision

18. How can human factors be designed into staff/management relations? (HF p. 336)

  a. Includes:

    - Behavior reinforcement

    - Job satisfaction, enlargement and enrichment

    - Motivation

    - Work effects on emotional stress

  b. Teaching human factors applications requires cooperation with trust between staff and management

19. How does human factors fit into accident investigation? (HF 337-338)

  a. Human Factors expertise is a recognized essential component

b.   Human Factors experts in all investigation teams

c.   Airlines should have in their safety programs

d.   Australian Bureau of Air Safety Investigation (BASI)

         -       Has a degree-qualified Human factors specialist

e.   There is still discussion about the need for human factors

f.   Public pressures, insurance and legal sources are pushing for it

20.   What part does human factors expertise have in the regulatory authority?   (HF p. 338)

       a.   Formerly national regulatory authorities had no international basis for requiring
                human factors knowledge as an element in training

       b.   In 1986 ICAO set the foundation for a program for aviation human factors

       c.   Not usual for the regulatory authority to employ a human factors specialist

21.   How important is an understanding of human factors to marketing?   (HF p. 339)

       -        Understanding of human behavior is of paramount importance

PO: 8.   Describe some of the difficulties in "selling" the ideas of using human factors,
               and list some of the benefits which can come from a future with better
               applications.

22.   Why must the presentation techniques and material used to describe human factors
       concepts employ the standards of human factors?   (HF p. 339)

       -        To prevent the loss of credibility

23.   Where do the origins of resistance to human factors progress often lie?
                                                                             (HF pp. 339-340)
       a.   Inflexible and out-dated attitudes and managerial concepts

       b.   Need for more formal management education in the technology

24.   How are the benefits of human factors applications best quantified?   (HF p. 340)

a.   Difficult to quantify benefits except in general terms only

b.   Cost of past human factor failures ----> Possible future Costs

c.   Rely on enlightened attitudes and policies rather than accounting methods

25.   What are some of the benefits of better application of human factors in aircraft operations?   (HF pp. 340-341)

a.   Higher quality tools and products

b.   More effective communication and documentation

c.   Better quality of leadership

d.   Reduced human error

e.   Reduced consequences of human error

f.   Minimizing of environmental effects on personal well-being

g.   Favorable attitude modification and stronger positive   motivation

h.   Increased level of job satisfaction

i.   Improved working efficiency

j.   Increased passenger satisfaction

k.   Improved cost effectiveness of training

26.   What is the key to success in human factors?   (HF p. 341)

-    Education is the key to success in Human Factors

EDUCATION AND APPLICATION

Human Factors   Human Factors
   Research   --------------------->   Applications

   Acidemia   Industry

   Knowledge   ----------------------> Practical Problems

Degree Qualifications:

- Most common background

    - Engineering, industrial or experimental psychology

    - A variety of psychology-based academic courses

QUALIFICATIONS

^^

Level 4 University higher degree ---->   Consultant
                 +
       Industry research

^^

Level 3 University degree   ---->   In-house
                                           Specialist

^^

Level 2 Two Week   ---->   Supervisors
       Compressed course

^^

Level 1 Awareness course   --->   All staff

^^

KLM ROYAL DUTCH AIRLINES

HUMAN FACTORS AWARENESS COURSE

Criteria:

- Awareness level coverage

- Maximum staff coverage at minimum cost

- Hardware mobility

- High credibility from human factors design quality

- Format and content - suitable for use by all staff

Course design:

- 15 Separate 40 minute self-contained audio-visual units

- Uses air transport examples to illustrate principles

SPECIFIC HUMAN FACTORS TRAINING

- Designed to meet a particular need within an operational environment

TEAM TRAINING

Line-Oriented Flight Training (LOFT)

Crew Resource Management (CRM)

LINE-ORIENTED FLIGHT TRAINING
(LOFT)

- Training and testing of flight crews as a team rather than as individuals

- Skills must be practiced as a group activity

- Personality and situational factors interact

- Particularly oriented towards the L-L interface

- Video-taping of the training session for self assessment

CREW RESOURCE MANAGEMENT

- The management and utilization of all

    - People

    - Equipment

    - Information available on the aircraft

- A special case of SHEL system management

- Generated from two assumptions:

    1- Accidents occur because the resources available are not adequately managed

    2- Resources are available to prevent accidents if they are optimally utilized

CRM ELEMENTS INCLUDE

- Leadership

- Communication

- Task distribution

- Setting priorities

- Monitoring

  - Information sources
  - Individual performance

- Extraction and utilization of data

CRM WILL NOT

- Change personality traits

- Change the effect on performance of:

  - Domestic stress
  - Fatigue
  - Circadian rhythm disturbance
  - Drugs or medication
  - Errors from information processing
  - Visual illusions
  - Design induced errors

HUMAN FACTORS APPLICATIONS

- Aircraft hardware

- Training

- Operating procedures and documentation

- Staff/Management relations

- Accident and incident investigation

- The regulatory authority

- Marketing

- Aircraft hardware

Design

Designer   Manufacturer   Operator   Certifying
                                         Authority

-    Six stages where Human Factors input must be made:

1-   System planning

2-   Pre-design

3-   Detail design

4-   Production

5-   Test/Evaluation

6-   Operations

-    Training

   -    Largely an operator responsibility

   -    Certifying authority - approves

   -    Manufacturer -   gives initial training

   -    Represents a major cost item

- Operating procedures and documentation

  - Basic operating procedures

    - Designed by the manufacturer

    - Approved by certifying authority

  - Primary responsibility is the operator under certifying authority supervision

- Staff/Management relations

  - Teaching human factors applications requires cooperation with trust between staff and management

- Accident and incident investigation

  - Human factors expertise is a recognized essential component

  - Human factors experts in all investigation teams

  - Airlines should have in their safety programs

  - Australian Bureau of Air Safety Investigation (BASI)

    - Has a degree-qualified human factors specialist

- The regulatory authority

  - Not usual for the FAA to employ human factors specialist

- Marketing

  - Understanding of human behavior is of paramount importance

QUANTIFYING HUMAN FACTORS BENEFITS

-     Difficult to quantify benefits

Costs of past
human factor   Possible future
failures   ---->   costs

-     Rely on enlightened attitudes and policies rather
      than accounting methods

BENEFITS OF BETTER APPLICATION
OF
HUMAN FACTORS

a. Higher quality tools and products

b. More effective communication and documentation

c. Better quality of leadership

d. Reduced human error

e. Reduced consequences of human error

f. Minimizing of environmental effects

g. Favorable attitude modification

h. Increased level of job satisfaction

i. Improved working efficiency

j. Increased passenger satisfaction

k. Improved cost effectiveness of training

-   Education is the key to success in human factors

Human Factors Investigation

The following section of human factors information (pages 345-351) is taken from the *US Air Force Guide To Mishap Investigation* (AFP 127-1 Vol I, Chapter 10, "Aeromedical Investigation", May 1987)

The same information is available to the students in the *Student Workbook,* pages 60-66.

The US Air Force source provides some additional insight on human factors issues and an investigative approach to the application of human factors. This information also provides an excellent summary of the main principles presented in the Hawkins text *Human Factors in Flight.*

The following diagram presents an overview of the areas of human factors which must be investigated for possible cause in a typical military aircraft mishap.

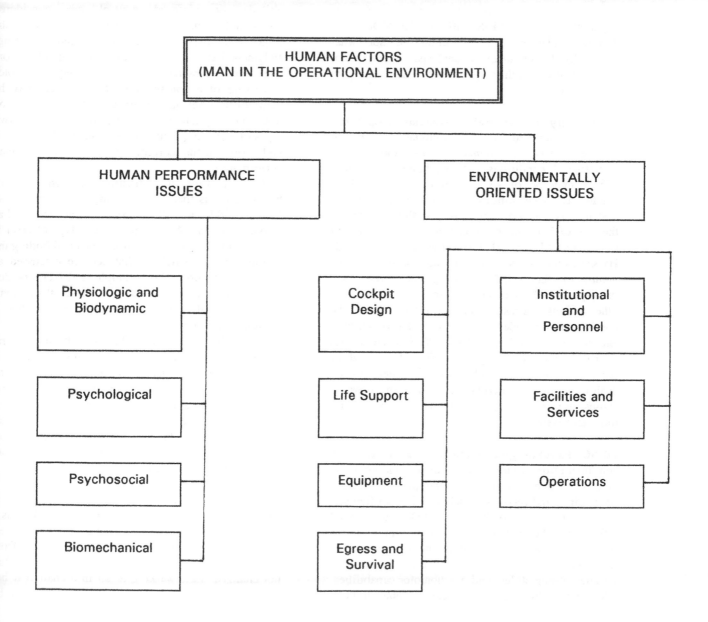

10-54. Human Performance Concerns. For purposes of both time management on the part of board members and completeness of topic and problem area coverage, the major areas of human factors interest are divided into man-centered human performance and environmental concerns. Human performance concerns are the traditional domain of the medical officer and his or her medical or psychological consultants, but must relate to environmental concerns addressed by other investigators. The issues considered during the medical data-consolidation phase should provide the basis for interaction with other board members who are integrating an overall picture of the mishap sequence during the team consultation phase. At this point, the need for specialized medical or psychological board augmentation may have been identified. As soon as a major issue in any human factors area becomes apparent, call HQ AFISC/SEL (AUTOVON 876-3458). The goal is to ensure a somewhat standardized and adequate analysis. This is not expected of the flight surgeon alone. The issues are often complex.

10-55. Physiologic and Biodynamic Concerns. These consider cardiorespiratory limitations, pressure change effects, the human senses, and various pathological conditions. A psychological training officer may be of some help in assessing these conditions. Acceleration effects on the cardiovascular system may include, among others, those related to the risk of inadequate blood flow to the brain, and those related to alveolar hypoxia under G. Hyperventilation is common, and because of the nature of the anti-G straining maneuver, may predispose to loss of consciousness along with its other effects. Pressure change effects should be considered to include hypoxia, evolved gas disorders, and various trapped gas effects. Various conditions of pathologic physiology as reflected in medical records, toxicology results, witness statements on health and fitness, thermal stress, and any possible causes for sudden incapacitation are basic and important issues.

10-56. Psychological Concerns. There must be continual efforts to improve our understanding of the factors which impact perception, information processing, and response. Behavior is not often self explanatory, and so it is essential that detailed information be gathered whether or not it seems pertinent initially. General problem areas include training, perception, attention, perceived stresses, fatigue, coping styles, and psychomotor capabilities. Although preliminary assessment of some of these

areas can be made by the flight surgeon, a team analysis of these areas is required. Only then can a meaningful integration of the different problem areas be achieved. When it is apparent that a psychologist would be useful to the board, there are some general criteria to consider in selection. An outline of these is provided in figure 10-22.

a. The concept of error pattern evaluation does not involve character judgement. For a multitude of reasons, simple human reality is that some behaviors are inappropriate to the task at hand. To intervene effectively, the specific associations must be teased out and studied. In essence, an error is seen from the standpoint of the competent investigator not as a cause so much as a manifestation of complex interaction between determinants of behavior. Technical errors such as missed radio calls or poor altitude or airspeed control are considered commonplace in aviation. Their reduction is the target of a variety of forms of proficiency training. Judgmental errors, on the other hand, involve more cognitive processing with the consequent choice consisting of a course of subtasks which may be inappropriate but only because they are a follow on to an inappropriate decision or judgment. These two types of errors are well known in safety, but are now undergoing further scrutiny by researchers in human performance.

(1) One type of distinction between types of human error is that between "slips" and "mistakes" (Morris and Rouse, 1985). Slips are characterized as errors of action. An example of this type of error is reaching for an automobile turn signal and finding the window wiper switch. Mistakes are discussed as errors of intention. For example, a driver stranded by a broken car with a history of electrical problems may replace the ignition computer rather than a broken fuel pump.

(2) Slips are thought to have several characteristics. They occur during some well rehearsed or established routine, appear to be associated with distraction or preoccupation, and flourish in familiar environments with few departures from the expected. Slips seem to be the result of actions that are highly automatic and so not consciously monitored. As a result, the expert may be even more prone to slips.

(a) The first and most common type of slip is where a habit pattern interference occurs. This is a response set where there is a change in routine (i.e., failing to stop at the store on the way home from work), a change in environment where the routine has not changed (i.e., walking to sit in a chair that has been removed), or the behavior is influenced by

346

environmental features (i.e., putting on one's coat instead of getting the box off the closet shelf).

(b) Unusual or ambiguous situations may facilitate a similar habit pattern phenomenon, a perceptual set. The perceptual set is the input side of this equation. A reading from an instrument may be that expected. Frequency is a factor when uncommon objects are misperceived as common ones. Incongruity, where an object does not "belong" in a setting may lead to a failure of perception, as can psychic need (i.e., a hungry person may perceive an ambiguous object as food).

(c) Omission or repetition of steps in intended sequences of events are "place losing errors." They are the result of systematization turned to ritualization (i.e., the use of a preflight checklist, missing, repeating, or not recalling a step).

(d) Slips can become distractions when they are recognized, and so perpetuate the problem. They can be expected to occur when environmental cues are not relevant to a present intention, when the environment has changed (cockpit configuration) but the task has not, when environmental features have not changed to facilitate an intended change in routine, when a long series of simple actions are required to complete a task, when the time between related actions is long or interrupted by other activity, or where procedures to do different tasks are similar in certain parts.

(3) Mistakes are more involved with judgment and decision making. Apparently well documented issues here are that decision makers consider no more than 2 or 3 variables at a time, and that recall may be triggered by prominent but irrelevant environmental cues. Attempts at solution may strive to incorporate the irrelevant information into the decision.

(a) The influence of past successes is probably disproportionately large and missing pieces of information are filled in based on individual theories which may have become reality to the decision maker. Once an operating hypotheses is selected, the natural tendency is to seek confirming evidence and explain away contradictions, often somewhat over-confident in one's state of knowledge.

(b) As a result of these characteristics of mistakes judgement, we can make some predictions. Conditions in which they are likely are: when more than two or three variables must be simultaneously considered, when prominent environmental cues suggest an inappropriate solution, when a particular inappropriate solution has been associated with success in past similar situations, and when choice of a solution requires approaching the problem in a novel fashion.

(4) Workload in this context is not viewed as a cause of error, but rather as a catalyst to it. In other words, an increase in workload may not necessarily lead to error, but will be far more likely to be associated with one when other conditions are conducive to it. The increase in workload may be seen as a stimulus to improved performance on one hand and a distraction on the other. Whether one views a task as motivational or distressing will have considerable influence. Thus, issues of morale and motivation come into view. The occurrence of both mistakes and slips compound subjective mental workload. Both dealing with the consequences of an error (or correction) and knowing one has committed an error would weigh in.

(5) It is best to request consultant assistance when significant workload questions arise.

b. When considering aircrew training, a distinction must be made between the quality of the program providing it and the knowledge level of the recipient at some later point in time. Learning ability, rate, transfer, and practice or rehearsal are interrelated. An understanding of memory, including immediate (working), short term, and long term must be used in assessing a response set or habit pattern. A pilot's skill and knowledge must be compared to what is provided in terms of procedural guidance and training programs (be they ongoing local proficiency, upgrade, or initial) to draw meaningful conclusions. A training inadequacy as a categorical finding deals only with a program. Other findings may relate to the individual. Remember that one-time exposure to information does not equate to knowledge or training adequacy. Expert and novice differences and the impact of automaticity should be considered. It may be useful to compare any findings here with what may be expected of comparable aircrew (not necessarily even in the same squadron, but certainly peers in mission). Survey aids can be more reliable than random personal interviews, but should be used carefully. See the Safety Investigation Workbook (volume III) for a sample model questionnaire and contact HQ AFISC to request assistance. The pilot or investigating safety board member input will be vital here.

c. Perception here is intended to refer to reading of sensory information rather than sensation itself. Cognition is here the process of integrating various sensory and internal cues. Cognitive flexibility will facilitate insight and efficiency and may increase with training and experience. Training will generally serve to reduce the cognitive or conscious information processing time required to accomplish a given task. It may create a "mindset," in this case a perceptual set that "expects" a certain environmental cue. A response set (a response essentially out of habit) is a related concept but at the outflow end of the decision process. Confusion may result from a breakdown of effective cognitive processes and it may lead to serious misperceptions. Cognitive saturation is, on the other hand, the result of the capacity for cognition being exceeded by the number of available pertinent cues. It is useful in part to think of cognitive or attentional resources as a single capacity entity. Yet, it is clear that the cognitive resources must be interpreted in light of stage of process (early vs. late), modality of processing (auditory vs. visual, central visual vs. peripheral, proprioceptive, etc.), processing codes (spatial vs. verbal), and attitudinal and stressor effects. This is by no means a complete picture as knowledge in the area grows daily.

d. Attention involves the mental process of directing cognitive resources. There is a limit to attentional or cognitive resources which varies among individuals as well as in an individual based upon the day or the situation. A focus of attention will consume some of this which leaves a margin of attentional reserve. When there is no reserve, there is cognitive saturation. Distraction, whether from the consciousness or the environment, interferes with attention. Fascination is seen when attention is arrested during a crisis situation. The result may be "shock" or a "freeze" behavior. Temporal distortion may be associated with a high stress situation. During high stress, time may be perceived as moving much more rapidly or slowly than is actual. Awareness of time is distorted. Channelized attention occurs when attentional resources are focused on a limited number of environmental cues of subjectively high priority. Inattention is, on the other hand, the failure to focus attention appropriately. It is here that the difficult problem of repetitive task effects (such as boredom and complacency) comes into view. The general form of inattention is associated with boredom or complacency. The selective form of inattention, on the other hand, results from lack of knowledge or an inappropriate set of expectations. It is apparent from this brief review that more commonly discussed issues such as unrecognized or Type I spatial disorientation must be evaluated in conjunction with these topics.

e. Perceived stresses are influential but in an individual sense. Expectations an individual holds regarding his or her environment are important. Consequences one believes are contingent upon performance (written or not) will guide one's behavior. When actuality falls short of expectations or when positive expectations go unfilled over a long period, distress may be experienced. Emotion, insight, perceived expectations of peers, supervisors or family, confidence in one's capability to deal with a situation, and perceived general workload all are powerful influences upon stress experienced as distress. The general adaptation syndrome (Selye) offers three phases in considering stress response. First is the fight or flight response which is immediate and temporary. Next, during the stage of resistance (the common stage), coping reserves are being actively directed toward adaptation in some form. Since this coping effort is active, it leaves diminished reserve. The final stage is that of exhaustion when the finite coping resources have been exceeded and symptoms my reappear.

f. Fatigue cannot be isolated from considerations of stress. Cumulative performance decrements can be a result of functioning in the stage of resistance over a long period. Acute performance decrements may be the result of high physiological and mental stress without adequate rest over some shorter period. Physical fatigue generally considers musculoskeletal limits of endurance. Sleep deprivation refers to an acute disruption of rest habits for whatever reason. Circadian cycles have a definite impact and must be considered based on "home" time. Motivational exhaustion refers to the emotional or affective component of fatigue and has a great deal to so with perceived stress experienced as distress. Biorhythms in the popularized good day/bad day form are not considered a proven entity. However, there may be a reason why stress is increased during weekend or holiday duty. Fatigue effects are pervasive, diminishing efficiency of mental processes from perception through exercise of judgment. Quantification can, however, be difficult.

g. Coping styles help to describe how an individual meets environmental demands. Decision making which includes the exercise of judgment and the selection of a response is further considered here. The actual mental process occurring is variable from one person to the next and from one time to the next. Studies have shown that what and individual describes about his or her decision is not necessarily accurate in reflecting what cognitive processing may have taken place. It may not be the same over time. Mental modeling is a complex issue. Inconsistencies may have nothing to do with integrity. This area will be a subject of continued research. It is important to keep this in mind in dealing with a witness or survivor interview.

(1) Personal discipline, general self confidence, motivation and other personality variables may play a role. Personality characteristics should be assessed but interpretation of this type of information must be

both candid and in consciousness of the biases of the observers. Such data may contribute to an explanation of why, for example, and individual over committed himself to a task.

(2) On the other hand, generalizations based on personality traits gathered in association with mishaps that might guide Air Force selection or other policies are unlikely.

h. Similarly, a subjective assessment of an individual's psychomotor capability may lead one to suspect a problem concerning strength or timing with control application. This, of course, must be done in light of what is "normal." Standards of normal may be broad and hard to apply even if available. Reaction time, for example, requires perception, diagnosis, exercise of judgment, selection of a response, and execution of the response. The time will vary depending on perceptual expectations, prior experience at exercising similar types of judgment, prior knowledge of alternative responses, practice experience (consequent skill) at response execution, and various factors that may diminish the proportion of cognitive resources that must be directed toward a given task. The time required to properly identify a situation and produce an unpracticed response may be as long as 6- to 9-seconds. This may have critical consequences.

10-57. Psychosocial Concerns. Here is where other supporting roles in the mishap may be exposed. Personal or community factors, supervisory influences, peer influences, and communication are among these concerns. The job environment is thought to be the more important area of focus. Because literature suggests this is a more proximate determinant of job behavior, the data may be more reliable, and some more direct intervention may be considered. (See paragraph 10-56 for comments on assistance.)

a. A person's perceived position within the community is again a matter of his or her expectations. These expectations of the environment in which the individual functions may not be easily assessed, but clues may be based on background such as education, travel, hobbies, religion, and career plans. Job satisfaction may be related to these, and to the extent to which the individual internalizes the values of the organization he or she purposes to serve. A powerful influence in this area may be close friends and family.

b. Supervisory issues are significant both psychosocially and institutionally. Command and control staff may have a powerful influence on behavior both by directives and enforcement of discipline, and by modeling (behavior that sets an

example). When an individual has been directly tasked to meet a standard, the pressure is expressed. However, it is perhaps even more important to recognize the numerous and powerful influences exerted by supervisors by their behavior (verbally and physically).

c. Peer influences are even more heavily weighted toward learning vicariously (by observation). What happens to one aircrew will be closely scrutinized by bright and observant fellows. Verbal peer comments only partly in jest may constitute powerful influences on behavior.

d. Communication concerns include personal habits in communicating, intracockpit information exchange, information exchange beyond the cockpit, and communication equipment failure. Cockpit resource management is a term that describes the pilot as a manager of all his or her resources. This becomes far more significant in a multicrew aircraft where task delegation is accomplished by communication. As a result, personal habits in communicating (including message generation, intonation, and listening) become critical. What behavior is professional and effective in crew coordination should be addressed both from the viewpoint of the aircraft commander and others in the cockpit. For any aircrew, the quality of information gained by communication with outside agencies can be vital. This may also be a concern in an environment where cluttering and confusion on the airwaves interferes and may also impact interaction between flight members.

10-58. Ergonomic and Biomechanical Concerns. These are an area of traditionally intense effort on the part of the medical member, and have been discussed as a part of autopsy consideration. Based on team investigative progress, there may occasionally be a need to return to x-rays to find evidence of sabotage, to photographs to interpret man-cockpit contact, or to toxicology to confirm the influence of cockpit smoke or fumes. As a result, the need for careful early management of perishable evidence is reconfirmed. Body habitus, size, and strength should also be evaluated where it is practical to do so. MIL-STD-1472C provides body size measurement data (and depicts the points of measurement) by fifth to ninety-fifth percentile for men and women.

10-59. Environmental Concerns. The other safety board team members are each experts in their own right. As a result, exchange between each of them and the flight surgeon may cover these environmental concerns in light of the physiologic or biodynamic, psychological, phychosocial, and anthropometric or biomechanical information garnered. This section is an outline of the concerns each of those board members address as a part of their safety or human factors analysis. The Safety Investigation Workbook (volume III) includes a summary sheet designed as a convenience aid to recall factors discussed with other board members when the time comes to assemble an overall report. The general concerns for each key word factor are presence, contribution, degree of contribution, and temporal role.

a. General topical concern areas are broken down into specific problem concerns. Key words within a problem or subproblem area may have somewhat overlapping meanings. Again, an attempt has been made to place them in an 11 topic hierarchy in a fashion that will make it easier to understand the various relationships. The terminology may undergo some evolution, but a glossary is included in the Safety Investigation Workbook.

b. The process of board consultation outlined here will facilitate a comprehensive and coherent analysis of pertinent human factors. Other consultants may have been called in for special problems as well. Integrating these various perspectives and extracting conclusions can improve reliability. Credibility of recommendations for either immediate measures (as generated by the medical board member), or those as a result of data trending over a number of mishaps, depends on the quality of investigation by the medical member and his or her fellow board member consultants. As a result, current references and consultants should be used freely to supplement the outline provided by this guidance. Again, call HQ AFISC/SEL (AUTOVON 876-3458) whenever an area of human factors concern is identified.

10-60. Cockpit Design.
This is a difficult challenge in small, multirole high-performance aircraft and in heavily automated multiplace cockpits. The problems may be weapons system specific and, as a result, the pilot member who is current in the mishap-type aircraft should review with the fight surgeon such problems as seat position, visibility, instrumentation, automatic systems and switch and control location. The idea is to assess the possibility for physical task saturation or to identify "designed in" impediments or limitations on mission accomplishment. (Accurate information is very important to designers.) Examples of problems in this area are numerous. The heads up display has been seen as a new problem area. The Instrument Flight Center at Randolph AFB TX is commissioned to study some of these problems, Aeronautical Science Division at Wright Patterson AFB OH Among others, also researches these issues.

10-61. Operations Concerns.
Flight-specific concerns should also be covered with the pilot member (and perhaps others). Mission demands begin with planning and briefing. Special flight stresses, such as range operations, aerobatic confidence maneuvers, various air combat tactics, and the possibility of acceleration displacement effects, should be more pertinent to fighter types. Special navigation, NAVAID, fatigue, or automation problems may well be more pertinent to larger aircraft. Also to be addressed are weather or night problems, emergencies (and pertinent emergency procedures), potential toxic exposures unique to an aircraft, and potential incockpit trauma.

10-62. Life Support and Personal Equipment.
Adequacy of this equipment to meet the special demands and risks of flight have been much improved through the mishap analysis of life-support officers and flight surgeons. This is a historic model for consultation to the flight surgeon. Both individuals discuss the cockpit environmental control and oxygen delivery systems, anti-G or pressure suit equipment, helmets, special mission gear (such as CBW), and other items of personal clothing and equipment. The life-support officer, when present, is normally a nonvoting board member.

10-63. Facilities and Services.
They should be discussed to include quality, availability, and any relationship to the mishap. The investigating safety officer will be a consultant in this area. The use of facilities is more a psychological or psychosocial concern, and access should also be considered here. Access to adequate nutrition, quarters for rest, and exercise, recreation, and health care facilities should be reviewed for any potential influence on a mishap sequence. In a more direct way, the facilities of an airfield or base, such as field lighting, weather service, aircrew dispatch, special intelligence, rescue or fire control services, or air traffic control, may also play a role.

10-64. Equipment Concerns.
Consider at the level of both local maintenance and that of the logistic system. The flight surgeon will assist the maintenance officer in evaluating local human concerns affecting maintenance. Many of the physiological, psychological, psychosocial, and even some of the anthropometric concerns enumerated may apply to the maintenance specialist and his or her supervisors. The important basic categories in which these principles may apply are evaluation of field quality assurance

and field working conditions (including tools and facilities). In addition, unit manning and individual qualification may be important. The maintenance officer may have additional questions concerning logistical considerations he or she must address, such as depot quality assurance, depot management, acquisition or modification philosophy, overhaul philosophy, or design defiency. However, these latter topics are the domains of systems safety.

10-65. Institutional or Management Issues. This concerns policies that may be shown to have had some relationship to a mishap. The experience and expertise of the board president make him or her a valuable consultant with whom to review issues of selection, evaluation, promotion, additional duties, the military or locally unique lifestyle, and internalization of unit and organizational values. Other board members may also be consulted, but it is appropriate to discard random speculation. Professional military education exposure and perhaps carefully configured surveys may be of some use.

10-66. Egress and Survival. Address all of the various man-centered concerns in the time frame after the point of the mishap. This point is the time when the mishap becomes inevitable regardless of crewmember action. In an ejection seat aircraft, there may be a need to assess the timeliness of an escape decision. Once again, the life-support officer is the consultant.

HUMAN FACTORS IN AVIATION SAFETY
(SF 320)
LECTURE NOTES

DAY 25   Human Factors Investigation

PO: *Describe and give examples of the Human Performance and Environmentally Oriented issues of human factors investigation as described in AFP 127-1 VOL I.*

AIR FORCE ACCIDENT INVESTIGATION (AFP 127-1 VOL I)

AEROMEDICAL INVESTIGATION (CH 10)

Section E - Human Factors Team Analysis

1. With what does human factors deal?   (IG p. 354 or SW p. 60)

    -        The man in the operational environment

2.   What are the two main issues to consider when investigating human factors?
                                                         (para. 10-54)

    1-   Human Performance issues

    2-   Environmentally Oriented issues

3.   What are the Human Performance issues of US Air Force human factors?
                                                         (paras. 10-55 - 10-58)

    -        Physiologic and Biodynamic

    -        Psychological

    -        Psychosocial

    -        Biomechanical

4.   What are some of examples of the physiologic and biodynamic human factors concerns?   (para. 10-55)

    -        Cardiorespiratory limitations: -   Acceleration effects

352

-	Pressure change effects:	-	Hyperventilation
		-	Hypoxia
		-	Evolved gas disorders
		-	Trapped gas effects
-	Human senses:	-	Visual or hearing impairment
-	Pathological conditions:	-	Toxicologic
			- Health and fitness
			- Thermal stress

5. What are some examples of the psychological human factors general problem areas?

(para. 10-56)

- Training

- Perception

- Attention

- Perceived stress

- Fatigue

- Coping styles

- Psychomotor capabilities

6. How is an error viewed from the human factors investigation standpoint?

(para. 10-56 a.)

- Not as a cause

- It is a manifestation of a complex interaction between determinants of behavior

7. What is the difference between technical errors and judgmental errors?

(para. 10-56 a.)

- Technical   = Missed radio calls etc

- Judgment   = Involve more cognitive processing

8. What is the difference between a slip and a mistake?

(para 10-56 a.(1)-(4))

- Slip   =   Characterized as errors of action

                      Occur during some well rehearsed or established routine

                      Associated with distraction or preoccupation

- Mistake   =   Errors of intention

                      More involved with judgment and decision making

                      Occur when more than two or three variables must be simultaneously considered

                      Issues of morale and motivation come into view

9. What is cognition?   (para. 10-56 c.)

- The process of integrating various sensory and internal cues

10. What is attention?   (para. 10-56 d.)

- The mental process of directing cognitive resources

11. What is fascination?   (para. 10-56 d.)

- When attention is arrested during a crisis situation

12. What are the psychosocial concerns the human factors investigator must examine?

(para. 10-57)

- Personal or community factors

- Supervisory influences

- Peer influences

- Communication

13. What human factors in the ergonomic and biomechanical areas must the safety investigators be concerned with?

(para. 10-58)

- Man-cockpit contact

- Cockpit smoke and fumes toxicology

- Body strength and size

14. What are the "Environmentally Oriented" issues of human factors investigation?

(para. 10-59)

- Cockpit Design

- Operations concerns

- Life support and Personal Equipment

- Facilities and services

- Equipment concerns

- Institutional, Personnel, and Management

- Egress and Survival

15. What human factors of cockpit design must be examined in the safety investigation?

(para. 10-60)

- Seat position

- Visibility

- Instrumentation

- Switch and control location

16. What are the operations concerns of human factors the safety investigator must check on?

(para. 10-61)

- Mission demands

- Special flight stresses

- Various air combat tactics

17. What life support and personal equipment aspects of human factors does the safety investigator need to examine?

(para. 10-62)

- Cockpit environmental control

- Oxygen delivery system

- Anti-G or pressure suit equipment

- Helmets

- Special mission gear (CW)

- Other items of personal clothing or equipment

18. Under facilities and services, what human factors must the safety investigator look into?

(para. 10-63)

- Access to adequate nutrition

- Quarters for rest

- Facilities for exercise, recreation, and health care

- Airfield facilities: - Airfield lighting

- Weather services

- Aircrew dispatch

19. What equipment concerns of human factors must be evaluated by the safety investigator?

(para. 10-64)

- Local maintenance: - Field quality assurance

- Field working conditions

- Unit manning and individual qualification

- Logistics system: - Depot Quality assurance

- Depot management

-       Acquisition or modification philosophy

20.   What are the institutional, training or management issues that can have an impact on
        human factors?

                                                                          (para. 10-65)
        -       Policies and issues of: -   Selection

                                              -       Evaluation

                                              -       Promotion

                                              -       Additional duties

                                              -       Internalization of unit and   organizational
                                                      values

21.   What are the egress and survival concerns of the human factors safety investigator?
                                                                          (para. 10-66)
        -       Assessing the timeliness of an escape decision

        -       Assessing aircraft crashworthiness for comparison with human impact
                tolerances

        -       Problems of survival: -   Heat

                                              -       Cold

                                              -       Water

                                              -       First aid

HUMAN FACTORS INVESTIGATION

THREE MAIN AREAS
OF
INVESTIGATION

1. The **Man**   -   Human Factors

2. The **Machine**   -   Airworthiness

3. The **Mission**   -   Operations

TWO MAIN HUMAN FACTORS ISSUES

1-   Human Performance issues

2-   Environmentally Oriented issues

HUMAN PERFORMANCE ISSUES

- Physiologic and Biodynamic

- Psychological

- Psychosocial

- Biomechanical

PHYSIOLOGIC AND BIODYNAMIC CONCERNS

- Cardiorespiratory limitations:

    - Acceleration effects

- Pressure change effects:

    - Hyperventilation

    - Hypoxia

    - Evolved gas disorders

    - Trapped gas effects

- Human senses:

    - Visual or hearing impairment

- Pathological conditions:

    - Toxicologic

    - Health and fitness

    - Thermal stress

PSYCHOLOGICAL GENERAL PROBLEM AREAS

- Training

- Perception

- Attention

- Perceived stress

- Fatigue

- Coping styles

- Psychomotor capabilities

ERRORS

- Technical = Missed radio calls etc

- Judgment = Involve more cogitative processing

- Slip - Characterized as errors of action

  - Occur during some well rehearsed or established routine

  - Associated with distraction or preoccupation

- Mistake - Errors of intention

  - More involved with judgment and decision making

  - When more than two or three variables must be simultaneously considered

  - Morale and motivation come into view

PSYCHOLOGICAL FACTORS DEFINITIONS

Cognition:

- The process of integrating various sensory and internal cues

Attention:

- The mental process of directing cognitive resources

Fascination:

- When attention is arrested during a crisis situation

PSYCHOSOCIAL CONCERNS

- Personal or community factors

- Supervisory influences

- Peer influences

- Communication

ERGONOMIC AND BIOMECHANICAL AREAS

- Man-cockpit contact

- Cockpit smoke and fumes toxicology

- Body strength and size

ENVIRONMENTALLY ORIENTED ISSUES

- Cockpit Design

- Operations concerns

- Life support and Personal Equipment

- Facilities and services

- Equipment concerns

- Institutional, Personnel, and Management

- Egress and Survival

COCKPIT DESIGN FACTORS

- Seat position

- Visibility

- Instrumentation

- Switch and control location

OPERATIONS CONCERNS

- Mission demands

- Special flight stresses

- Various air combat tactics

LIFE SUPPORT AND PERSONAL EQUIPMENT ASPECTS

- Cockpit environmental control

- Oxygen delivery system

- Anti-G or pressure suit equipment

- Helmets

- Special mission gear (CW)

- Other items of personal clothing or equipment

FACILITIES AND SERVICES

- Access to adequate nutrition

- Quarters for rest

- Facilities for exercise, recreation, and health care

- Airfield facilities:

    - Airfield lighting

    - Weather services

    - Aircrew dispatch

EQUIPMENT CONCERNS

- Local maintenance:

    - Field quality assurance

    - Field working conditions

    - Unit manning and individual qualification

- Logistics system:

    - Depot Quality assurance

    - Depot management

    - Acquisition or modification philosophy

INSTITUTIONAL, TRAINING
OR MANAGEMENT ISSUES

- Policies and issues of:

    - Selection

    - Evaluation

    - Promotion

    - Additional duties

    - Internalization of unit and organizational values

EGRESS AND SURVIVAL CONCERNS

- Assessing the timeliness of an escape decision

- Assessing aircraft crashworthiness for comparison with human impact tolerances

- Problems of survival:

  - Heat

  - Cold

  - Water

  - First aid

Video Involvement Questions # 1

"ALERT 3" (Crash of United Flight 232)

This is a video recording of a presentation made by Capt Al Haynes to a group at the NASA Ames Research Center, Dryden Flight Research Center at Edwards AFB, CA on 24 May 1991. He discusses the crash of United Flight 232 at Sioux City, Iowa on the 19th of July 1989.

The introduction is a sound track of some of the ATC communications with United 232, then Capt Haynes makes his presentation.

PO: 1.   *Answer the questions designed to highlight various details of the video presentation.*

PO: 2.   *Identify the aspects of human factors that are found in the presentation given by Capt Haines on the crash of United 232. Use the SHEL conceptual model to describe good and poor interface examples.*

Sound Track

1.   How was the crew controlling the turns in their DC-10?

      a.   ailerons only
      b.   rudder only
c   c.   power only
      d.   a combination of a & c

2.   Which direction of turns did the crew tell the controller they could make?

a   a. Right   b. Left   c. Left or right

3.   Initially did the crew think they would make it to the airport?

c   a. Yes   b. No   c. There were serious doubts

4.   What was the final result of the minor misunderstanding about elevator authority?

      a.   They did have elevator control
      b.   They had minimal elevator control
c   c.   They had no elevator control
      d.   They had intermittent elevator control

5.   What directions did the crew give to the ground emergency response vehicles?

      a.   Place them at the approach end of the runway.

b   b.   Place them at the departure end of the runway.

      c.   Locate them at mid field

      d.   Put them in the corn field at the end of the runway.

6.   How far were they from the airport when they put the landing gear down?

b   a. At 30 mi   b. Outside 10 mi   c. Between 5-10 mi. d. Inside 5 mi

7.   What runway did tower clear United 232 to land on?

d   a. RWY 22   b. RWY 35   c. RWY 17   d. Any runway

Capt Haynes' presentation

8.   What were the five factors Capt Haynes said made it possible for them to have as many survivors as they did?

      (1) _____   (Luck)

      (2) _____   (Communications)

      (3) _____   (Preparation)

      (4) _____   (Execution)

      (5) _____   (Cooperation)

Factor # 1

9.   What did he say about getting the airplane on the ground with only two engines for control?

(It happened over the relative flat lands of Iowa - The best possible place)

10.   How did he feel about the weather and turbulence at the time of his emergency?

(No turbulence, on thunder storms - good weather)

11.   What part did the time-of-day play in the emergency?

(Daylight - at shift change time for the hospitals)

12. How did the Air National Guard fit into factor # 1?

(The Air Guard was on duty with 285 trained personnel)

Factor # 2

13. What did Capt Haynes say about the controller who gave them directions to the field?

(Used a very calm voice and was always giving vital information)

14. What did he say about radio frequency congestion after they declared an emergency?

(Lots of help is available when you declare an emergency)

15. How did Capt Haynes feel about the cockpit and cabin crew communications during the emergency?

(It was good although not much was used with the cabin crew)

16. What results did factor # 2 have on the emergency response forces?

(15 minutes notice that an airplane has crashed)

Factor # 3

17. How did factor # 3 effect the response of the crash/rescue and other participants in the disaster?

(They had a plan. A practice drill identified discrepancies that had been corrected)

18. What was the key to the success of the response of the cabin crew?

(Flight attendant simulator training)

19. What cockpit crew training did Capt Haynes say was the most valuable for them?

    a.   Simulator training in total hydraulic failure situations.
c   b.   Cockpit Resource Management (CRM)
    c.   Line Orientated Flight Training (LOFT)

Factor # 4

20. What did the crew discover would minimize the pitch oscillations?

(Do just the opposite of what it would seem)

21. What response error did Capt Haynes say the crew was making in the attempt to keep the wings level?

(Over-controlling the power differential)

22. Who ended up working the throttles?

    a.   Capt Haynes
    b.   The 1st officer
    c.   The Flight Engineer
d   d.   An extra DC-10 Capt

Factor # 5

23. How much combined experience operating the DC-10 did the cooperation of each of the cockpit crew total?

c   a. 65 years   b. 90 years   c. 103 years   d. 115 years

24. What insightful way were the passengers prepared to increase the chance of survival for the children?

(At least one adult next to each child)

25. How many people from the surrounding areas did Capt Haynes say lined up to offer blood ?

(400)

Video Involvement Questions # 2

"WHY PLANES CRASH"

This video is a NOVA presentation, written, produced, and directed by Veronica L. Young. Executive producer -Paula S. Aspell in 1987. Coronet Films Video, 108 Wilmont Rd., Deerfield, IL 60015.

PO: 1. *Answer the questions designed to highlight various details of the video presentation.*

PO: 2. *Describe the aspects of human factors that are found in the video presentation "Why Planes Crash." Use the SHEL conceptual model to describe good and poor interface examples.*

1. How often do airline passengers think of the pilot or crew being at fault in their natural concern for their safety?

c a. Frequently b. About half the time c. Rarely

2. How many people fly on planes in a year?

d a. 100 million b. 500 million c. 700 million d. 900 million

3. How many people are killed world wide in plane accidents in a year?

a a. 800 b. 1000 c. 1500 d. 2800

4. What are the three things that are happening which pose a safety challenge for aviation?

 1. recovery from the _(air traffic controller's strike)_

 2. _(Experience factor)_ - new pilots and controllers in the system

 3. deregulation environment - more _(airlines)_

5. What does Capt Roger Brooks say about why the margins of safety are reduced?

 - _(Operating philosophy)_ has changed based upon increased _(competitive)_ pressures

6. What did John Nance say lead to the Air Florida Flight 90 accident in 1982?

 - _(A chain of human failures)_

7.   What was the first human failure noted that caused the Air Florida 737 to not have enough power to continue the climb after liftoff?

- The crew did not _(turn on engine anti-ice)_

8.   What other deicing procedures were also ignored?

- _(Taking off with ice and snow on the wings)_

9.   What is the range of accidents blamed on cockpit error?

- _(60 - 80%)_

10.   What responsibility did the first officer have in the Air Florida crash?

- He knew something was wrong but _(was not forceful enough in expressing)_ himself.

11.   What side of the "safety equation" did Capt Mel Volz say needed attention now?

- the _(human)_ side

12.   Where are the roots of some of the human factors problems on the flight decks of today?

- in the macho _(self sufficiency)_ of the single seat fighter pilot

13.   What were some of the motivational feelings given for being a pilot?

- feeling of exhilaration, _(accomplishment)_ and _(mastery)_

14.   What does Dr Clay Foushee, the aviation psychologist, say might be parts of today's "right stuff" in the cockpit?

a. concern with the _(task)_ (technical skills)

b. ability to _(relate)_ well to other _(people)_

15.   What are the two aims of Cockpit Resource Management?

a. modify _(behavior)_ to a degree

b. get the crew to _(work together as a crew)_

16. What does Dr Clay Foushee see as one of the most useful techniques in CRM training?

- the full mission simulation with _(video replay)_ of actions

17. In CRM deficiency evaluation what, in addition to how well the crew handled the technical challenges, is also evaluated?

      a.   How well the emergency ended

b   b.   Their coordinated effort

      c.   How precisely they controlled the aircraft

      d.   How responsive the ATC system was

18. What is one of the values coming from the "video feedback" in CRM?

- allows one to see how _(they come across)_ as a crew member

19. What recurring problems are illustrated by the disaster of Eastern Flight 401 which crashed int he Florida Everglades?

- failure of captains to act as leaders, _(make decisions)_ ,

set _(priorities)_ , and _(delegate)_ responsibilities

- shortcomings which are compounded by _(unassertive)_ and _(complacent)_ crew members

20. What did the captain neglect to do which allowed everyone else on the crew to be absorbed in the crisis?

      a.   Declare an emergency

      b.   Communicate the nature of the problem

c   c.   Divide up flying responsibilities

      d.   All of the above

21. What was wrong with the controllers inquiry "Eastern 401, How are things coming along out there"?

      a.   It was too vague

      b.   It didn't communicate to the crew the real concern

      c.   It didn't alert the crew to the altitude deviation

d   d.   All of the above

22. What was one reason given by a first officer for not making mention to the captain of an observed problem?

        a.   The captain could not understand the language.

        b.   There was too much noise in the cockpit.

c   c.   The captain would really "jump" on him.

-   d.   The captain would not listen anyway.

23. What does Dr John Lauber say his point of view is about CRM training?

        a.   It is a refined method of training that is proven to be very effective.

b   b.   It is a concept still in development.

        c.   The training concept has received universal acceptance in the airlines.

        d.   Both a and c were stated.

24. Which of the airlines (at the time of this video) had full fledged CRM training?

a&b   a. Pan Am   b. United   c. Delta   d. Southwest

f&g   e. Eastern   f. Peoples Express   g. Continental Express

25. According to Dr Clay Foushee why are accidents a terrible research criterion for evaluating success of any program?

a   a.   because they are so infrequent

        b.   because there are so many fatalities with each one

        c.   because they are so frequent

        d.   because human factors are so hard to determine

26. Why does Del Fadden say the pilot will not be eliminated from the flight deck in the foreseeable future?

- The pilots skill at _(cognitive)_ _(decision)_ _(making)_ can not be duplicated

27. How does the Boeing 757 and 767 relieve the pilots' mental workload?

- by using a computer which _(augments)_ decision making

28. How does the computer system of the glass cockpit save mental effort of the pilot in decision making?

- by _(assembling)_ the _(information)_ needed for a _(decision)_

29.   What does Dr Earl Wiener say are the two "D" problems of any automatic device?

     a.   Delicate and dangerous
     b.   Dumb and dangerous
     c.   Dutiful and delicate
d   d.   Dumb and dutiful

30.   What is one of the human factors problems of monitoring automation?

    - _(complacency)_____ or lack of vigilance

31.   What does Dr Wiener say is a very realistic concern the pilots have about the increasing use of automation in the cockpit?

     a.   They will loose their jobs
b   b.   The will have a loss of skills
     c.   The computer will make an unrecoverable error
     d.   Computers just can't fly as well as the human pilots

32.   What does John Nance say the problems in the aviation industry are based on?

    - lack of _(management (understanding)_____ of how close you can come

    to the line and not _(decrease)_____ _(safety)_____

Video Involvement Questions # 3

"THE WRONG STUFF"

This video is a HORIZON presentation, written and produced by Jeremy Taylor. Film editor - Peter Essex. HORIZON editor Robin Brightwell

PO: 1.   *Answer the questions designed to highlight various details of the video presentation.*

PO: 2.   *Point out the aspects of human factors that are found in the video presentation "The Wrong Stuff." Use the SHEL conceptual model to describe good and poor interface examples.*

1.   What does the video "The Wrong Stuff" focus on?

a   a.   Flight crew behavior
        b.   Failures of the ATC system
        c.   Problems of new technology
        d.   Modern weather problems

2.   What are the two very important facts about air safety that come from examining crashes?

        1.   Commercial aviation is still incredibly _(safe)_____.

        2.   Accidents _(still)_____ happen and in about _(80%)_____ of them the pilot is to blame.

3.   In the famous incident of the 747 at Nairobi cited by Roger Green, why did the crew choose not to believe the warnings about being below the glide slope?

        a.   Visual cues verified the warnings were false
        b.   There was disagreement among the crew
        c.   The crew was distracted by an engine failure
d   d.   They did not fit into their model

4.   According to Dr John Lauber's review of the 707 crash into the mountain on Barla Indonesia while attempting an NDB approach, what did the crew fail to do?

        a.   recognize station passage
        b.   set in the correct altimeter setting
c   c.   positively resolve the ambiguities
        d.   verify an obviously wrong assigned altitude by ATC

5. In the Boeing 727 simulation with a No 3 engine fire, what error in the cockpit resource management did the captain make relative to the flight engineer?

a. gave him the wrong configuration
b
b. overloaded him with tasks
c. did not accept his vital inputs
d. sent him back in the cabin to fight the fire

6. According to Dr Bob Helmreich which of the 3 components at the test pilot right stuff creates problems when trying to function as an effective team?

a. High technical competence
b. Rugged individualism
c. High level of competitiveness
d
d. Both b and c above

7. What seems to be the ironic result from the lack of copilot assertiveness illustrated by accidents like the Air Florida crash?

a
a. they would rather die than stick their necks out
b. they become over assertive captains
c. they can't read engine instruments as well as captains
d. their flying skills are better than most captains

8. When captains in a simulator study pretended to be incapacitated, what percent of the simulators crashed when the copilots failed to take over?

d
a. 10%   b. 15%   c. 20%   d. 25%

9. As a result of the United 707 crash at Portland how did the airline rethink its approach to pilot training?

a. It is basically a _(man)_____ _(management)_____ problem.

10. In the United CRM training discussion how is the 9-9 management style defined?

a. advocates his position
b. inquires from the rest of the crew
c. concerned with accomplishing the task
d. concerned with input of the crew
e. follows through on conflict resolution
f. can make a decision
g
g. all of the above

11. In the review of the video replay of the simulator flight for the United CRM training, what did the instructor point out was a significant thing being done?

      a.   the captain was dominating the situation
      b.   the crew was focusing too much on the emergency
c   c.   there was a good division of responsibilities
      d.   the copilot was not as assertive as he should have been

12. What did Peoples Express focus on to eliminate the wrong stuff from their cockpits?

      a.   extensive use of cockpit resource management
b   b.   selection screening for people who can work with people
      c.   line of flight training (LOFT) for flight and cabin crew
      d.   selection of good looking people with good technical skills

13. What has drastically changed the nature of the pilot's job in modern transport aircraft?

a   a.   computer automation
      b.   higher performance engines
      c.   wide bodied aircraft
      d.   using only 2 pilot crews

14. What does the "Electronic Cocoon" do to help strike the balance between human and machine?

      a.   allows the crew more freedom to operate
      b.   provides warning to the crew when they approach limits
      c.   creates a dialogue between the human and machine
d   d.   all of the above

15. Why does Dr John Lauber say the computer will never replace the pilot on the flight deck?

a   a.   only people can be creative
      b.   people can fly as good as computers
      c.   people will never trust computers totally
      d.   it's better to have a captain than a computer first on the accident scene

Video Involvement Questions # 4

"TOP GUN AND BEYOND"

This video is a NOVA presentation, written, produced and directed by Chris Haws.
Executive Editor - William Grant. Executive producer - Paula S. Aspell

PO: 1.   Answer the questions designed to highlight various details of the video presentation.

PO: 2.   Discuss the points of human factors that are found in the video presentation "Top Gun and Beyond." Use the SHEL conceptual model to describe good and poor interface examples.

1.   What are the engineers of today's fighter aircraft not adequately dealing   with?

c   a. Psychology   b. Technology   c. Biology   d. Statistics

2.   At the speeds of today's modern fighter aircraft how much time is available for a fighter pilot to deal with the enemy who is 20 miles away?

b   a. 15 sec   b. 30 sec   c. 1 min   d. 2 min

Next there will be about a 15 minute historical development of fighter aircraft technology.

Make note of any human factors considerations indicated during this period.

3.   What was the result of all the various inputs the pilots of the Vietnam era had to deal with?

a   a.   Pilot saturation and overload
       b.   Cockpit display design improvements
       c.   Peak pilot psychological performance
       d.   The best designed aircraft ever developed

4.   What happens to a fighter pilot if he does not unload and get the blood back in his head after experiencing tunnel vision and first blackout?

       a.   He sees black and white dots.
       b.   He experiences "red out".
c   c.   He experiences a loss of consciousness.
       d.   He has heart failure.

5. What is the estimate of pilot deaths in 5 years from "G" loss of consciousness?

d   a. 5   b. 10   c. 15   d. 20

6. How long does it take for the higher learning, cognitive centers of the brain to start working adequately after a "G" loss of consciousness incident?

c   a. 12 sec   b. 30 sec   c. 2 min   d. 10 min

7. How many extra "Gs" of tolerance can the anti "G" straining maneuver bring the pilot?

b   a. 1   b. 3   c. 5   d. 8

08. What new study is being conducted to help the aircraft monitor the status of the pilot?

a   a.   Brain wave monitoring
      b.   Blood pressure monitoring
      c.   Eye movement monitoring
      d.   Breathing monitoring

09. What are the "smart systems" being developed at Wright-Patterson AFB designed to help the pilot with?

      a.   elevated brain blood pressure
      b.   better instrument scan patterns
c   c.   greater situational awareness
      d.   improved aircraft control inputs

10. With all the technological developments, why can't the computer replace the pilot?

a   a.   the human flexibility advantage
      b.   the computer visual deficiency
      c.   the computer lack of processing speed
      d.   the lack of computer accuracy

APPENDIX:

MASTER QUESTION FILE

To prevent the compromise of the test questions, this Master Question File should be stored, in a controlled access area separate from the *Instructor's Guide*.

INTRODUCTION

This Appendix is designed to help you measure the students achievement of the concepts in the *Human Factors in Flight* textbook. All the questions are either multiple choice, matching or true/false and are arranged by the unit which coincides with the *Instructor's Guide* and *Student Workbook* units. They are organized into three test groups for ease in developing examinations. Usually each concept has two or more questions associated with it. This provides an instructor the ability to select alternate versions of the test where questions are different but the same student objectives are measured.

Table of Contents

TEST # 1

The number following the unit title is the number of questions needed from that unit to build a 50 question test # 1 performance objectives.

Background to Human Errors   5

01. What percent of the accidents in the final approach and landing phase were charged to the flight crew as a primary factor in the Boeing report on the <u>worldwide commercial jet</u> fleet?

b   a. 87.7%   b. 78.9%   c. 58.2%   d. 39.3%

02. How does "Personnel" fit in as a broad cause factor in the 1988 <u>general aviation</u> aircraft accidents?

a   a. 87.7%   b. 78.9%   c. 58.2%   d. 39.3%

03. In the Boeing report, what percent of the accidents in the past 10 years in the worldwide jet fleet have flightcrew as a factor in <u>all</u> <u>phases</u> of the flight profile?

d   a. 87.7%   b. 78.9%   c. 58.2%   d. 39.3%

04. From the NTSB Annual Review of U.S. Air Carrier accidents, what was the percent given to "pilot" as a broad cause or factor in the five year period 1983-1987?

c   a. 87.7%   b. 78.9%   c. 58.2%   d. 39.3%

05. Which of the following best describes the Human Factors milestone accomplished at the Psychology Laboratory at Cambridge?

     a.   Effectiveness could be favorably influenced by psychological factors not directly related to the work itself.
b   b.   Skilled behavior was dependent to a considerable extent on the design, layout and interpretation of displays and controls.
     c.   Establishment of ergonomics or Human Factors as a technology in the founding of the Ergonomics Research Society in the UK in 1949.
     d.   In 1982 the UK set up the Confidential Human Factors Incident Reporting Programme (CHIRP).

06. Which of the following best describes the Human Factors milestone accomplished at the International Air Transport Association (IATA) 20th technical conference at Istanbul in 1975?

     a.   Effectiveness could be favorably influenced by psychological factors not directly related to the work itself.
     b.   Skilled behavior was dependent to a considerable extent on the design, layout and interpretation of displays and controls.
c   c.   The turning point in official recognition of the importance of Human Factors in air transportation.
     d.   Establishment of the first "Human Factors Awareness Course" for large scale, low-cost, in-house indoctrination of staff in basic Human Factors principles.

07. When did serious interest in generating a greater awareness of human factors among those responsible for design, certification and operation get started?

c   a. 1950s   b. 1960s   c. 1970s   d. 1980s

08. How is Human Factors different from Ergonomics?

     a.   Ergonomics is the broader concept.
     b.   Ergonomics deals more in the environmental aspects while Human Factors concentrates more on the human.
     c.   Human Factors centers more on the physiological while ergonomics deals more with psychological.
d   d.   Human Factors encompasses some aspects of human performance and systems interfaces not generally considered in ergonomics.

09. Human factors is about **people** in their working and living environment. It also is about **people** in their relationships with machines, procedures, the environment and other people.

a   a. True   b. False

10.   Human factors is sometimes identified with a branch of medicine because of the earlier human factors problems being physiological.

a   a. True   b. False

The SHEL Conceptual Model   7

Match the appropriate components of the SHEL conceptual model listed below (a -d) with the example of the interface described in the examples cited in items 01 - 04. (Some may be used more than once)

       a.   Software   (L-S)
       b.   Hardware   (L-H)
       c.   Environment   (L-E)
       d.   Liveware   (L-L)

01.   Designing displays to match the information processing characteristics of the human.
b
02.   Considering the problems associated with disturbed biological rhythms and sleep deprivation.
c
03.   Designing procedures, manual and checklist layout, symbology and computer programs to match human processing patterns.
a
04.   Using Line of Flight Training (LOFT) to enhance leadership, crew coordination, teamwork and personality interactions.
d
05.   Fitting flyers with helmets against noise, flying suits against cold, oxygen masks against the effects of altitude, and Anti-G suits against acceleration forces.
c

06.   Which of the below listed Liveware component characteristics is best described by a concern with long and short term memory, and the effects of stress and motivation?

       a.   Physical size and shape
       b.   Input characteristics
       c.   Output characteristics
       d.   Individual differences
e   e.   Information processing

07.   Which of the below listed Liveware component characteristics is best described by a concern with body measurements and movement?

a   a.   Physical size and shape
       b.   Input characteristics
       c.   Output characteristics
       d.   Environmental tolerances
       e.   Information processing

08.   Which of the below listed Liveware component characteristics is best described by a concern for sensing and information needed to enable response to external events?

       a.   Physical size and shape
b   b.   Input characteristics
       c.   Output characteristics
       d.   Individual differences
       e.   Information processing

09.   Which of the below listed Liveware component characteristics is best described by a concern for the kind and direction of forces for the movement of the controls?

       a.   Physical size and shape
       b.   Input characteristics
c   c.   Output characteristics
       d.   Environmental tolerances
       e.   Information processing

10. Which of the below listed Liveware component characteristics is controlled by selection, training, and application of standardized procedures?

    a. Physical size and shape
    b. Input characteristics
    c. Output characteristics
d   d. Individual differences
    e. Information processing

The Nature of Error   6

01. What characteristics do errors that are induced by poorly designed equipment or procedures have?

    a. They are likely to be repeated.
    b. They are largely predictable.
    c. We must look to the designer not the operator for correction.
    d. a and c only.
e   e. a, b, and c are all true.

02. What things did the Three-Mile Island nuclear incident, the Chernobyl nuclear disaster, and the Tenerife double 747 crash have in common?

    a. They came from human error occurring in a working environment deficient in Human Factors
    b. Serious warnings of inadequate Human Factors and possible devastating consequences were given before the accidents
    c. Effective measures to heed the warnings were not taken
    d. Only a and b above
e   e. All of the above (a, b & c)

03. What are the two basic tenets with respect to human error?

    1. The origins of error can be fundamentally different.
    2. The origins of error are fundamentally the same.
    3. The consequences of similar errors can be quite different.
    4. The consequences of similar errors are fundamentally the same.
    5. Most human errors are based in the psychological.
    6. All human errors have the same basic origin.

a   a. 1 & 3   b. 2 & 4   c. 5 & 6   d. 2 & 3   e. 1 & 5

04. Which of the following is **NOT** true of the use of the term "pilot error"?

a   a. The term leads to more in-depth mishap prevention efforts.
    b. It implies the nature of error made by this operator is unique.
    c. It has obstructed progress toward greater flight safety.
    d. The concept focuses more on who happened rather on why it happened.
    e. It has impeded a more profound and rational examination of human performance.

05. "Pilot errors" are different in principle from errors made by anyone else.

b   a. True   b. False

06. If human error is normal, of what value is the human operator?

    a. The human is a very flexible system component.
    b. If ergonomics has been properly applied to system design, the human can give increased overall system reliability.
    c. As system design becomes more complex the value of the human is less important.
d   d. Only a and b are proper expressions of the value of the human operator.
    e. Only b and c are proper expressions of the value of the human operator.

07. Which of the following are expressions of normal human error rates?

    a. Simple, repetitive tasks are 1 in 100
    b. 1 in 1000 is good in most circumstances
    c. The rate varies widely depending on factors such as the nature of the task or risk involved, motivation and sleep loss and fatigue.
d   d. All of the above

08.   What does "a tendency by some people to have more accidents than others with equivalent risk
       exposure, for reasons beyond chance alone" define?

       a.   High risk tasking
b   b.   Accident proneness
       c.   Low motivation
       d.   Simple carelessness

09.   What seems to have the greatest effect on the short term factor of simple carelessness?

a   a.   Changes in motivation
       b.   Repetitive training
       c.   Variations in rewards expected
       d.   State of stress or fatigue

10.   How does the short term influence of stress affect the human?

       a.   It usually increases ability to perform.
       b.   It usually decreases ability to perform.
c   c.   It varies from one person to another.
       d.   It has little effect on performance. Only long term stress effects human performance.

Sources of Error   5

01.   Which of the following best describes an error caused by a mismatch in the Liveware-Software
       interface?

       a.   Poorly designed warning systems
       b.   An inappropriate authority relationship between the captain and the first officer
c   c.   Deficiencies in conceptual aspects of a warning system
       d.   Reduced performance caused by disruption of biological rhythms

02.   Which of the following best describes an error caused by a mismatch in the Liveware-Hardware
       interface?

a   a.   Poorly designed warning systems
       b.   An inappropriate authority relationship between the captain and the first officer
       c.   Deficiencies in conceptual aspects of a warning system
       d.   Reduced performance caused by disruption of biological rhythms

03.   Which of the following best describes an error caused by a mismatch in the Liveware-Environment
       interface?

       a.   Poorly designed warning systems
       b.   An inappropriate authority relationship between the captain and the first officer
       c.   Deficiencies in conceptual aspects of a warning system
d   d.   Reduced performance caused by disruption of biological rhythms

04.   Which of the following best describes an error caused by a mismatch in the Liveware-Liveware
       interface?

       a.   Poorly designed warning systems
b   b.   An inappropriate authority relationship between the captain and the first officer
       c.   Deficiencies in conceptual aspects of a warning system
       d.   Reduced performance caused by disruption of biological rhythms

05.   Which of the areas of the information processing system of the liveware component handles the
       conclusion reached about the nature of the message received?

b   a. Sensing   b. Perception   c. Decision-making   d. Action   e. Feedback

06.   Which of the areas of the information processing system of the liveware component would be most
       concerned with the deterioration from age or a physical disorder?

a   a. Sensing   b. Perception   c. Decision-making   d. Action   e. Feedback

391

07. Which of the areas of the information processing system of the liveware component would be concerned with distortions caused by emotional or commercial considerations?

    a.   Sensing
    b.   Perception
c   c.   Decision-making
    d.   Action
    e.   Feedback

08. Which of the areas of the information processing system of the liveware component would be concerned with adverse effects of badly designed controls?

d   a. Sensing   b. Perception   c. Decision-making   d. Action   e. Feedback

09. The optimum level of arousal for accuracy in performance is higher than that for speed.

b.   a. True   b. False

10. Which of the following statements about the human decision-making process is **NOT** true?

    a.   The most dangerous characteristic of the false hypothesis or mistaken assumption is that it is frequently extremely resistant to correction.
    b.   Man has a vast capacity for sensory input but only one channel for decision-making.
c   c.   Mistaken assumption is more likely to occur when attention is focused on the task rather than diverted elsewhere.
    d.   Following a period of high concentration or when expectancy is high, mistaken assumption is more likely.

Error Classification & Reduction   8

01. Which of the following errors often has only one or two factors and is easier to correct than the others?

b   a. Random   b. Systematic   c. Sporadic   d. Omission

02. Which of the following errors have no discernable pattern and may be influenced by many factors?

a   a. Random   b. Systematic   c. Sporadic   d. Reversible

03. Which of the following errors can occur after routinely good performance and are very difficult to predict?

c   a. Random   b. Systematic   c. Sporadic   d. Substitution

04. Which kind of error would be best described by failing to do something which ought to be done?

d   a. Random   b. Systematic   c. Commission   d. Omission

Match the examples listed in items 05 - 07 with the error classification (a - c) which **BEST** fits. (Use each only once)

    a. Commission   b. Omission   c. Substitution

05. Missing an item on the checklist
b
06. Shutting down the wrong engine with an engine fire
c
07. Silencing the gear warning horn during configuration
a

08. If you design a computer data entry control panel to verify or challenge an entry before the entry is actually input into the automatic flight control, in which of the below listed classification systems are you applying your principle for error reduction?

    a.   Design-induced and Operator-induced
    b.   Random, Systematic and Sporadic
    c.   Omission, Commission and Substitution
d   d.   Reversible and Irreversible

09.   From the lists of tasks below, which one is performed best by the human rather than the machine?

a   a. Error correction   b. Deductive reasoning   c. Complex activity   d. Monitoring

10.   Which of the following is **NOT** a way to minimize the occurrence of human error from the "vigilance decrement" phenomenon?

       a.   Assign the task to a machine
       b.   Space the tasking so as to not exceed 30 minutes of task demand
       c.   Transfer the activity to data link
d.   d.   Enhance selection or training practices

11.   Which of the following best fits as a way to describes the concept of reducing the consequences of errors?

       a.   Ensuring a high level of staff competence through optimum selection, training and checking
b   b.   Dispelling the illusion that it is possible to have "error-free" operation
       c.   Personality, attitudes and motivation play a vital role
       d.   Focus on the quality and condition of the L component of the system
       e.   Tolerance to fatigue and other stresses are important

12.   Which error classification designation has more impact in the concept of reducing the consequence of error?

       a.   Design-induced and Operator-induced
       b.   Random, Systematic and Sporadic
       c.   Omission, Commission and Substitution
d   d.   Reversible and Irreversible

13.   Optimum level of arousal depends on the nature of the task performed and a complex task requires **more** arousal than a simple task.

b   a. True   b. False

14.   In the context of the SHEL model, minimizing the occurrence of error means we must work with people as they are rather as we would like them to be.

a   a. True   b. False

Fatigue, Body Rhythms, and Sleep   13

01.   Which of the following causes of fatigue is the most probable to occur from "Jet Lag"?

       a.   Inadequate rest
b   b.   Disruption or displaced body rhythms
       c.   Excessive muscular or physical activity
       d.   Excessive cognitive activity

02.   Which of the following might be included in the dangers which come from flying with "jet lag"?

       a.   Slowed reaction and decision time
       b.   Defective memory for recent events
       c.   Tendency to accept lower standards of operational performance
       d.   Sleep disturbance and deprivation
e   e.   All of the above

03.   Which of the following statements about the nature of the problem of fatigue is **NOT** true?

       a.   93% of pilots report it as a problem.
       b.   NASA reported in 1981 that fatigue associated decrements in performance resulted in substantive potentially unsafe aviation conditions.
       c.   The British CHIRP system announced in 1984 that the largest number of reports received concerned fatigue, sleep and the way the work patterns are constructed
d   d.   Confidential reports show a surprising gain in knowledge and concern among pilots about human factors concerning fatigue.

04.   The Galvanic Skin Response (GSR) system used to assist in countering some of the effects of jet lag detects low arousal by monitoring eye movements.

b   a. True   b. False

393

05.   Which of the four causes for fatigue listed below was probably the most recent one discovered and seems to have the most direct effect on the flight crew?

       a.   Inadequate rest
       b.   Excessive muscular or physical activity
c   c.   Disruption or displaced body rhythms
       d.   Excessive cognitive activity

06.   Which of the following is most significant in regulating body rhythms?

a   a.   Rotation of the planet in 24 hours
       b.   Amino acids, cortisol and other hormone levels
       c.   International Society of Chronobiology
       d.   Sodium and potassium excretion

07.   Which of the following has the greatest effect on maintaining the human body's circadian rhythm?

a   a.   Entraining agents called zeitgebers
       b.   The 24 hour pacemaker located in the cerebral cortex of the brain
       c.   The sodium, potassium and amino acid excretion cycle
       d.   The "daily constitutional" in the bathroom
       e.   The clock radio on the night stand

08.   The term "acrophase" in chronobiology means the lowest point in the rhythm curve for each of the body's systems.

b   a. True   b. False

09.   In describing the biorhythm and the effect it has on human performance, which of the following statements is **NOT** true?

       a.   The curve is task-dependent and will vary according to the task.
       b.   Maximum and minimum performance scores within a cycle is task dependent and tends to be greater with complex than simple tasks.
       c.   Practice, motivation and increased effort will raise and flatten the performance curve.
d   d.   Loss of performance from loss of a single night's sleep may be greater than that resulting from being out of cycle with the natural circadian rhythm.
       e.   There is a post-lunch dip where we may want to avoid critical tasks which require optimum performance.

10.   Which of the following statements about the rhythm of performance in the 24 hour cycle is **NOT** true?

       a.   The curve is task-dependent and will vary according to the task.
b   b.   Maximum and minimum performance scores within a cycle is task dependent and tends to be lesser with complex than simple tasks.
       c.   Practice, motivation and increased effort will raise and flatten the curve.
       d.   Loss of performance from the natural cycle may be greater than that coming from loss of a single night's sleep

11.   When does the circadian rhythm of oral temperature peak out on the average person?

c   a. 1200 hrs   b. 1500 hrs   c. 1800 hrs   d. 2100 hrs

12.   Statistical analysis of a very large number of aircraft accidents has failed to establish any correlation between the three cycles of the Birthdate Biorhythm Theory and the timing of aircraft accidents.

a   a. True   b. False

13.   Which of the following statements is **NOT** true concerning the resynchronisation which occurs following transmeridian travel?

       a.   Systems shift their phase at different rates that are out of phase with local time and out of phase with each other.
b   b.   Most travelers find recovery from westbound flights is harder than from eastbound flights.
       c.   Resynchronisation does not appear to have a constant rate.
       d.   There is substantial differences in the ability of individuals to adjust their circadian rhythms to repeated transmeridian shifts.

14. Which of the following statements about the disturbance of biological rhythms is most appropriate to use with regard to flight crews?

      a. Circadian dysrhythmia
      b. Desynchronosis
c   c. Transmeridian dyschronism
      d. Metergic dyschronisim

15. Which of the following statements about the use of drugs to counter the effects of sleep loss or biorhythm disruption is **in error**?

a   a. Barbiturates are the least dangerous drugs for pilots to use to induce sleep.
      b. Benzodiazepines like Valium have an adverse effect on performance.
      c. Alcohol induces sleep but the pattern is not normal with a suppression of REM sleep.
      d. Caffeine disrupts the pattern of sleep, reducing Stage 4 and REM.

16. One of the most significant conclusions from research is that sleep appears to play a role in the maintenance of motivation.

a   a. True   b. False

17. There is a demand for a drug which would accelerate the resynchronisation.

a   a. True   b. False

18. No totally effective and acceptable drug has been developed for routine and long-term application in the control of the problems associated with transmeridian dyschronism.

a   a. True   b. False

19. Which of the following best describes the characteristics of paradoxical or REM sleep?

      a. Tensed muscles in the throat
      b. More frequent major body movements
c   c. Rapid eye movements
      d. Brain wave patterns are slow
      e. There is normally no dream recall

20. Which of the following best describes the characteristics of orthodox sleep?

      a. Tensed muscles in the throat
      b. More frequent major body movements
      c. Rapid eye movements
d   d. Brain wave patterns are slow
      e. There is normally no dream recall

21. Which of the following statements about sleep is **NOT** true?

      a. Normal sleep pattern disruption is the most common physiological symptom of long-range flying.
      b. REM sleep occurs about once every 90 minutes.
      c. During a normal night sleep shifts from one stage to another about 30 times.
      d. A nap must be not less than 10 minutes duration for it to be restorative.
e   e. There are 4 stages or levels to paradoxical sleep and the normal human shifts from one stage to another 30 times in a nights sleep.

22. Which of the following statements about sleep **IS** true?

      a. REM sleep occurs only in the nap or microsleep.
b   b. During a night normal sleep shifts from one stage to another about 30 times.
      c. A nap must be not less than 30 minutes duration for it to be restorative.
      d. There are 4 stages or levels to paradoxical sleep.

23. Which of the following correctly describe the effect sleep has on memory?

    a. There seems to be an increase in information retention just before dropping off to sleep.
    b. Sleep and night-time are better for memory than wakefulness and daytime.
    c. Restorative processes are enhanced due to the increase in the net rate of protein synthesis during sleep.

d   d. All of the above
    e. Only a and b above are correct.

24. Which of the following statements best describes situational insomnia?

    a. Difficulty sleeping under normal conditions in phase with body rhythms.
    b. Rarely a disorder itself but usually a symptom of another disorder.
    c. Difficulty in sleeping in a particular situation when biological rhythms are disturbed.
    d. Usually physiological and related to body chemistry.

e   e. Both c and d are descriptions appropriate to situational insomnia.

25. Which of the following best describes Autogenic Training?

    a. Incorporates a regular, vigorous exercise program followed by a cold shower.
    b. Involves a system of using the drug melatonin to train the body to accelerate the circadian rhythm adjustment.

c   c. Involves the use of passive concentration aimed at producing a state of psychophysiological relaxation.
    d. It is an easy to learned-it-yourself technique which facilitates enhancement of certain homeostatic, self-regulating mechanisms in the body.

26. How can exercise affect sleep?

    a. It increases the quality of sleep by increasing the REM.
b   b. It increases the slow wave sleep (SWS).
    c. It decreases the quality by decreasing the SWS.
    d. There is no proven effect that exercise has on sleep.

Fitness and Performance   7

01. Which of the following is the more frequent source of pilot incapacitation?

b   a. Heart attack   b. Gastrointestinal   c. Hypoglycemia   d. Appendicitis

02. Which of the following reasons for pilot incapacitation would be most likely to result in total incapacitation?

d   a. Fatigue   b. Stress   c. Medication   d. Stroke

03. Being fit is a condition which permits a generally high level of physical and mental performance.

a   a. True   b. False

04. A person who is fit has an ability to perform with minimal fatigue, to be tolerant to stress and to be readily able to cope with changes in the environment.

a   a. True   b. False

05. Which of the following are results of a good **physical** fitness program?

    a. Improved body systems
    b. Better mental performance
    c. Improved motivation and reduced tension
d   d. All of the above (a b & c)
    e. Only a & b above)

06. Which of the following are results of a good **physical** fitness program?

    a. Improved psychological state and psychomotor performance
    b. Resistance against fatigue and better physiological performance
    c. Improved self-esteem
d   d. All of the above (a b & c)
    e. Only a & b above)

07. Which of the following types of exercises fit the three types of exercise needed for a full fitness program?

      a.   Upper body strength, lower body strength, and endurance
      b.   Back, legs and upper body strength
c   c.   Mobility, heart/lung, and strengthening
      d.   Golf, leg lifts, and weight lifting

08. Which of the following activities BEST fits the three types of exercise needed for a full fitness program?

b   a. Walking   b. Swimming   c. Golf   d. Baseball

09. Which of the following statements about the effects of smoking is **NOT** true?

      a.   Nicotine is the source of satisfaction and addiction in smoking.
b   b.   The smoker's reaction time **decreases** when they are deprived of tobacco.
      c.   Although cigars and pipe tobacco produce more carbon monoxide than cigarettes, the levels in the blood are usually higher with cigarette smokers.
      d.   Aerobic performance of smokers is significantly worse than non-smokers.

10. Which of the following is **NOT** an effect of smoking?

      a.   Increases adrenaline and non adrenaline output
      b.   A raised level of physiological arousal
c   c.   Complex task performance is better
      d.   Deteriorates central nervous system (CNS) function
      e.   Long-term memory somewhat better

11. In the U.S. in 1971, the cost of alcoholism in industry was about $ 10 billion a year and the overall cost of alcohol abuse was estimated at about $ 25 billion a year.

a   a. True   b. False

12. One of the insidious flying performance implications for alcohol users which may not be generally known is that higher mental and reflex functions can be affected for two to three days after a "serious drinking session"

a   a. True   b. False

13. What percent of general aviation pilots killed in the U.S. were found to have a BAC of 15 mg% or higher?

b   a. 10%   b. 20%   c. 30%   d. 40%   e. 50%

14. What percent of the pilot population have used drugs on the job at any point in their flying career?

d   a. 21%   b. 25%   c. 32%   d. 46%   e. 58%

15. Which of the following is a **psychological** stress producer that reduces performance?

a   a.   The feeling of insecurity associated with the taking of a six-month medical or proficiency check.
      b.   The disturbing influence of transmeridian flying on the circadian rhythms.
      c.   The fatigue coming from the environmental noise, vibration, or temperature extremes.
      d.   Both b and c

16. Which of the following is a **physiological** stress producer that reduces performance?

      a.   The feeling of insecurity associated with the taking of a six-month medical or proficiency check.
      b.   Emotional and domestic stress from family separation
c   c.   The fatigue coming from the environmental noise, vibration, or temperature extremes.
      d.   Both b and c

17. Which of the following ways of managing the response individuals make to the stresses in their lives is the **LEAST** effective?

    a.   Autogenic Training
    b.   Advisory groups
    c.   Employee Assistance Programs (EAP)
    d.   A company sponsored Comprehensive Program

e   e.   Drug and alcohol therapy

18. What is the keystone of the comprehensive program to tackle latent or active personal stress conditions?

    a.   Personal counseling for the individual, stress response modification, and medical treatment for the individual.
    b.   Workplace examination, work organization analysis and modification, and personal affairs review conducted by the human factors specialist.

c   c.   The test which provides for each individual an analysis of his own psychological and biochemical stress profile.
    d.   The peer advisory group.

19. Which of the following best describes the human performance benefits from **fats** in the diet?

    a.   Reduces the conditions which bring on constipation, hemorrhoids, diabetes and certain forms of cancer.
    b.   Have no energy-producing value but are essential for maintaining health
    c.   Are needed for the building and repair of body tissues
    d.   Are absorbed rapidly and provide one of the chief and most immediate sources of energy

e   e.   Provide the most concentrated source of heat energy in the body

20. Which of the following best describes the human performance benefits from **proteins** in the diet?

    a.   Reduces the conditions which bring on constipation, hemorrhoids, diabetes and certain forms of cancer.
    b.   Have no energy-producing value but are essential for maintaining health

c   c.   Are needed for the building and repair of body tissues
    d.   Are absorbed rapidly and provide one of the chief and most immediate sources of energy
    e.   Provide the most concentrated source of heat energy in the body

22. Which of the following best describes the human performance benefits from **fiber** in the diet?

a   a.   Reduces the conditions which bring on constipation, hemorrhoids, diabetes and certain forms of cancer.
    b.   Have no energy-producing value but are essential for maintaining health
    c.   Are needed for the building and repair of body tissues
    d.   Are absorbed rapidly and provide one of the chief and most immediate sources of energy
    e.   Provide the most concentrated source of heat energy in the body

23. Which of the following best describes the human performance benefits from **carbohydrates** in the diet?

    a.   Reduces the conditions which bring on constipation, hemorrhoids, diabetes and certain forms of cancer.
    b.   Have no energy-producing value but are essential for maintaining health
    c.   Are needed for the building and repair of body tissues

d   d.   Are absorbed rapidly and provide one of the chief and most immediate sources of energy
    e.   Provide the most concentrated source of heat energy in the body

"Alert 3 - The Crash of United 232"   6

Questions 01 - 12 relate to the video "Alert 3 - The Crash of United 232" and the massages given by Capt Haynes.

01. How was the crew controlling the turns in their DC-10?

c   a. ailerons only   b. rudder only   c. power only   d. a combination of a & c

02. Initially did the crew think they would make it to the airport?

c   a. Yes   b. No   c. There were serious doubts

03.   What was the final result of the minor misunderstanding about elevator authority?

        a.   They did have elevator control
        b.   They had minimal elevator control
c   c.   They had no elevator control
        d.   They had intermittent elevator control

04.   What directions did the crew give to the ground emergency response vehicles?

        a.   Place them at the approach end of the runway.
b   b.   Place them at the departure end of the runway.
        c.   Locate them at mid field
        d.   Put them in the corn field at the end of the runway.

05.   What cockpit crew training did Capt Haynes say was the most valuable for them?

        a.   Simulator training in total hydraulic failure situations.
b   b.   Cockpit Resource Management (CRM)
        c.   Line Orientated Flight Training (LOFT)

06.   Who ended up working the throttles?

        a.   Capt Haynes
        b.   The 1st officer
        c.   The Flight Engineer
d   d.   An extra DC-10 Capt

07.   How much combined experience operating the DC-10 did the cooperation of each of the cockpit crew amount to?

c   a. 65 years   b. 90 years   c. 103 years   d. 115 years

Match the appropriate one of the five factors (a - e) which Capt Haynes said made it possible for them to have as many survivors as they did with the examples given in items 08 - 12. (Use each only once)

        a.   (1) Luck
        b.   (2) Communications
        c.   (3) Preparation
        d.   (4) Execution
        e.   (5) Cooperation

08.   Passengers followed instructions well and each child was teamed with an adult before the crash.
e
09.   The time the hospital was notified of the emergency was right at shift change and there was twice the medical staff on duty.
a
10.   The Sioux City Airport had a flexible emergency plan that was recently refined during a practice drill.
c
11.   The combined experience of the crew with the help of the extra DC 10 captain helped discover how to use the throttles to maneuver the aircraft.
d
12.   The ATC radar controller who directed them to the airfield was calm voiced and provided exceptional assistance.
b

TEST # 2

Vision   4

Match the following terms of light measurement (a-e) with their appropriate definition in items (01-05)
(use each only once)

     a. Contrast   **b**. Luminance   **c**. Refraction   **d**. Intensity   **e**. Illumination

01.   Brightness of the reflected light from the surface
b

02.   The relationship of brightness or luminance of a target to its background or surroundings
a

03.   Point source brightness
d

04.   Light falling on the surface
e

05.   The bending of light as it goes through a medium of different density
c

Match the following terms of the physical makeup of the eye (a-e) with their appropriate
definition in items (06-10)   (use each only once)

     a. Pupil   **b**. Retina   **c**. Fovea   **d**. Extrinsic muscle   **e**. Ciliary muscle

06.   Changes the shape of the lens to modify the focal length
e

07.   A complex layer of nerve cells at the back inside of the eye where the image is projected
b

08.   Controls the amount of light entering the eye
a

09.   The center of the retina where only cone receptors are located
c

10.   The six external muscles which control the movement of the eye
d

11.   What component of the human eye is **BEST** described with the term "Photopic vision"?

e   a. Retina   b. Cornea   c. Lens   d. Rods   e. Cones

12.   What component of the human eye is **BEST** described with the term "Scotopic vision"?

d   a. Retina   b. Cornea   c. Lens   d. Rods   e. Cones

Match the following functions of the eye (a-e) with their appropriate definition in items (13-16)
(use each only once)

     a. Stereopsis   **b**. Adaptation   **c**. Rhodopsin   **d**. Visual acuity   **e**. Mandelbaum effect

13.   The substance called visual purple in the rods which is bleached out in bright light
c

14.   Refers to the smallest letter which a subject can read on a chart at 20 ft.
d

15.   The effect provided by the eye's converging for depth perception
a

16.   Adjustment for various light conditions
b

17.   Which of the following is the more significant adjustment made by the eye during dark adaptation?

a   a.   The action of the light sensors in the retina using the pigment called rhodopsin or visual purple.
       b.   The course adjustment of the pupil to allow more light into the eye.
       c.   The shifting of the "blind spot" from stereoptic vision to central or fovea vision.
       d.   The lens thickening to increase refraction and place the center of focus in the scotopic vision area.

18.   What are the two basic adjustments of the eye needed to see clearly at different distances?

      1.   Switch from photopic vision to scotopic vision
      2.   Change the refractive power of the lens
      3.   Accommodation of the pupil size with the iris
      4.   Adjusting the extrinsic muscles for binocular vergence
      5.   Adjustment of the rods and cones for visual acuity

b   a. 1 & 2   b. 2 & 4   c. 2 & 3   d. 3 & 4   e. 3 & 5

19.   Which of the below listed terms is associated with the accommodation function of the eye?

      a.   Mandelbaum effect
      b.   Binocular vergence
      c.   Adaptation
      d.   Stereopsis
e   e.   Only a & b

20.   To what does "The effect provided by the eye's extrinsic muscle converging the eyes for depth perception" have reference?

a   a. Stereopsis   b. Adaptation   c. Binocular vergence   d. Accommodation

21.   Where in the perceptual process do uncertainty and ambiguity occur with correctly sensed information?

c   a. Eyes   b. Inner ear   c. Brain   d. Vestibular apparatus

22.   When perception is influenced by expectation it is sometimes called _____.

c   a. Fascination   b. Blind spot   c. Set   d. Critical fusion

23.   Which of the following statements is **NOT** true concerning the effect of smoking on vision?

a   a.   Visual acuity is enhanced by the nicotine
      b.   The brightness threshold and reaction to visual stimuli deteriorated
      c.   Individuals suffer differently in performance degradation
      d.   Depletion of oxygen supply to the brain occurs
      e.   The effects of altitude and smoking are cumulative

Visual Illusions   4

01.   Which term is used BEST in describing a visual illusion of movement created by an isolated stationary light in an otherwise dark visual field appearing to wander?

a   a. autokinesis   b. somatogravic   c. oculogravic   d. oculogyral

02.   Which term is used BEST in describing a visual illusion of movement created by a false sense of turning?

d   a. autokinesis   b. somatogravic   c. oculogravic   d. oculogyral

03.   Which term is used BEST in describing a **visual response** to the illusion of tumbling after moving your head while in a turning aircraft causing the fluid in the third semicircular canal to move from the motion of the other two?

a   a. Coriolis   b. Somatogravic   c. Oculogravic   d. Oculogyral

04.   Which term is used BEST in describing a **visual response** to the illusion of pitching up during acceleration?

c   a. Coriolis   b. Somatogyral   c. Oculogravic   d. Oculogyral

401

05. Which of the following approach conditions can cause an visual illusion that will **result** in the pilot flying too **close** to the ground?

1. Up sloping terrain to the runway	2. Down sloping terrain to the runway
3. An up sloping runway	4. A down sloping runway
5. A wider than normal runway	6. A narrower than normal runway
7. Dimmer approach and runway lights	8. Brighter approach and runway lights
9. Increased visibility conditions	10. Reduced visibility conditions

b   a. 2,4,6,8,&10   b. 1,3,6,7,&10   c. 1,3,5,7,&9   d. 2,4,5,8,&9   e. 1,3,5,7,&10

06. Which of the following approach conditions can cause a visual illusion that will **result** in the pilot flying too **high** from the ground?

1. Up sloping terrain to the runway	2. Down sloping terrain to the runway
3. An up sloping runway	4. A down sloping runway
5. A wider than normal runway	6. A narrower than normal runway
7. Dimmer approach and runway lights	8. Brighter approach and runway lights
9. Increased visibility conditions	10. Reduced visibility conditions

d   a. 2,4,6,8,&10   b. 1,3,6,7,&10   c. 1,3,5,7,&9   d. 2,4,5,8,&9   e. 1,3,5,7,&10

07. If your aircraft has a lower pitch angle from a faster speed, this will place the target aircraft lower in the windscreen and give the illusion that the aircraft is at a lower altitude.

b   a. True   b. False

08. If your aircraft has a higher pitch angle from a lower speed, this will place the target aircraft higher in the windscreen and give the illusion that the aircraft is at a higher altitude.

b   a. True   b. False

09. Which of the following is **NOT** correct in defining the effect of a visual illusion?

    a.   Distances tend to be overestimated in conditions of poor visibility.
    b.   A relative higher pilot eye height will give the illusion of reduced apparent relative motion.
c   c.   When the ground is sloping down towards the runway an impression is given of being too high.
    d.   There is an illusion of height when all is dark except for the distant runway or airport lights.

10. Which of the following is **NOT** true concerning the protective measures which can be used to guard against visual illusions?

a   a.   Concentrate mostly on the visual cues and don't supplement them with information from other sources.
    b.   Approach and airport charts where geographic locations are known to be associated with visual illusions should be so noted.
    c.   All concerned with flying must recognize that visual illusions are a normal phenomena.
    d.   We must understand their nature and situations in which they are likely to be encountered.

11. What does the "design eye" refer to?

    a.   The design of the cockpit instrument panel with the attitude indicator at the center
b   b.   The eye position in the cockpit where the pilot can see a specified length of the approach or touchdown lights
    c.   The required eye position for the lower surface of the glare shield to provide a lateral horizontal reference and also coincide with the forward downward vision angle
    d.   The design of the 20/20 eye which is not susceptible to visual illusions or perceptual distortions

Motivation and Safety   6

01. When did the National Transportation Safety Board (NTSB) first establish a separate Human Performance Division to look int the psychology of human motivation in aviation safety?

e   a. 1938   b. 1948   c. 1965   d. 1975   e. 1983

02.   What is probably the most significant characteristic of the Liveware component in driving a person to behave in a particular way?

b   a. Conditioning   b. Motivation   c. Training   d. Punishment

03.   The basic level of motivation begins as a need for _____.

b   a. belonging   b. survival   c. affection   d. recognition   e. self actualization

04.   What is a sequence of motivated behavior called?

a   a. Goal   .   b. Pattern   c. Drive   d. Instinct

Match the following Theories associated with motivation (a-c) with the appropriate statement in items 05 - 07. (Use each only once)

       a. Learning Theory   b. Psychoanalytic Theory   c. Cognitive Theories

05.   Explains a person's concept of self or ego as the drive behind motivation.
b
06.   Emphasizes the characteristics of man as a rational being which makes choices freely.
c
07.   Relates present motivation to a particular action which was rewarded in the past.
a

08.   In Maslow's hierarchy of needs the example of the basic level of motivation is centered in _____

       a.   an expression of capacities and talents.
       b.   a need for friendship, affection and love.
c   c.   the drive for obtaining oxygen, food and water.
       d.   achievement, prestige, status and dominance.
       e.   obtaining freedom from pain and danger.

09.   In Maslow's hierarchy of needs the example of the highest level of motivation is centered in _____

a   a.   an expression of capacities and talents.
       b.   a need for friendship, affection and love.
       c.   the drive for obtaining oxygen, food and water.
       d.   achievement, prestige, status and dominance.
       e.   obtaining freedom from pain and danger.

Match the below listed industrial studies (a-c) with the findings or theories which emerged from them as listed in items 10 - 12.   (Use each only once)

       a.   Taylor - 1890s at Bethlehem Steel Company
       b.   Hawthorne studies - 1920s at Western Electric
       c.   Two-Factory Theory of Herzberg - 1959

10.   Work was influenced by social and psychological factors quite independent of the work itself.
b
11.   Money was considered the prime motivating factor.
a
12.   Satisfaction came from motivating factors like achievement, recognition for good work, and responsibility.
c

13.   Which of the motives cited by Murray **BEST** fits with Maslow's "self-esteem, prestige, or status needs" hierarchy of needs?

b   a. Affiliation   b. Achievement   c. Power   d. Expectancy

14.   Which of the motives cited by Murray **BEST** fits with Maslow's "belonging and affection" hierarchy of needs?

a   a. Affiliation   b. Achievement   c. Power   d. Expectancy

15. Which of the below descriptions of Murray's motives best fits with his concept of **Achievement motivation**?

      a.   Involves adherence and loyalty to a friend

b   b.   Qualities that are paramount importance in jobs with a high level of unsupervised performance

      c.   Concern over the means of influencing the behavior of another person

      d.   Concerned with the establishment and maintenance of affectionate relationships

16. Which of the below descriptions of Murray's motives best fits with his concept of **Power motivation**?

      a.   Involves adherence and loyalty to a friend

      b.   Qualities that are paramount importance in jobs with a high level of unsupervised performance

c   c.   Concern over the means of influencing the behavior of another person

      d.   Concerned with the establishment and maintenance of affectionate relationships

17. Which of the below descriptions of Murray's motives best fits with his concept of **Affiliation motivation**?

a   a.   Involves adherence and loyalty to a friend

      b.   Qualities that are paramount importance in jobs with a high level of unsupervised performance

      c.   Concern over the means of influencing the behavior of another person

      d.   Drive towards achievement for its own sake rather than material benefits

Influencing Motivation   2

18. In the power of expectancy in motivation, what is the most important aspect of the availability of rewards?

      a.   That they are valued enough by management

      b.   What management says is available

c   c.   What the employee sees as available

      d.   That the worker's reward equal to or better than a peer's

19. Improvement in job satisfaction is not necessary for, nor does it automatically result in improved performance.

a   a. True   b. False

20. A rationally critical attitude of an employee towards the supervisory or management policies is a sure indication of job dissatisfaction or low motivation.

b   a. True   b. False

21. If you want to increase a person's motivation through **job enrichment** what must you do?

      a.   Give a pay raise

      b.   Increase the number and variety of tasks

c   c.   Involve the person in policy and decision-making

      d.   Provide more benefits such as recreational facilities or time off

22. Which of the following is **NOT** true concerning leadership in motivating work performance?

      a.   There is a powerful need to improve an understanding and application of the management of human resources on the flight deck.

b   b.   Authority is normally acquired while leadership is assigned and suggests a voluntary following.

      c.   One of the prime tasks of leadership is to encourage the desired motives in others in the group.

      d.   All members of the group contribute to the effective leadership of the group by supplying information, contributing ideas, providing support and by their general response to the leader.

23. Which of the following statements about leadership is **NOT** correct?

a   a.   Authority is acquired and suggests voluntary following

      b.   Optimally - authority is combined with true leadership

      c.   All members of a group contribute to the effective leadership of the group

      d.   The leader modifies habits and behavior by reinforcement

01.   Which of the elements of the communication process uses language?

        1- Sender   2- Message   3- Medium   4- Receiver

e   a. 1 & 4   b. 2 & 3   c. 1,2, & 4   d. 2,3, &4   e. 1,2,3, & 4

02.   Which of the following statements concerning ambiguity in the communication process is true?

        a.   It is a characteristic of communication which varies with the different form of language.
        b.   Sometimes speaking clarifies meaning of words with the same appearance.
        c.   Sometimes writing clarifies meaning of words sounding the same.
        d.   Sometimes context only can clarify the meaning of a word with the same sound and appearance.
e   e.   All of the above are true.

03.   Which of the following statements about the language of communication is **NOT** true?

        a.   Language is inextricably linked with the cognitive or thinking process as well as with
                communication.
        b.   Ambiguity is a characteristic of communication which varies with the different form of
                language.
        c.   The principles or rules which form the foundation for the arrangement of words in a language
                are   called syntax or grammar.
d   d.   The role of written and spoken language is the same.

04.   Which of the following BEST relates to the uniqueness of **spoken** language?

        a.   Has the characteristic of permanence in time and space
        b.   Good for information accumulation and storage
        c.   Uses the visual channel
d   d.   Uses pronunciation and accent for clarification

05.   Which of the following BEST relates to the uniqueness of **written** language?

a   a.   Uses punctuation and accent for clarification
        b.   Provides rapid exchange of messages
        c.   Primary lubricant of social interaction
        d.   Uses the auditory channel

06.   Concerning the **intelligibility** of a word in spoken language, generally, the **shorter** the word the more
        readily it is identified.

b   a. True   b. False

07.   Which of the following will have the greatest intelligibility of the various types of speech if the
        articulation is reduced?

        a.   A vocabulary limited to 1000 phonetically balanced words
        b.   A vocabulary limited to 256 phonetically balanced words
c   c.   A vocabulary limited to 52 phonetically balanced words
        d.   A vocabulary with words known to the listener

08.   The U.S. was the pioneer in the research which led to the standard international vocabulary and
        phraseology and leads the way for application.

b   a. True   b. False

09.   Which part of the human hearing apparatus is affected when a hearing loss results from **nerve deafness**?

b   a. Outer ear   b. Organ of Corti   c. Middle ear   d. The brain

10.   Which part of the human hearing apparatus is affected when a hearing loss results from **otosclerosis**?

c   a. Outer ear   b. Organ of Corti   c. Middle ear   d. The brain

11.   Which part of the human hearing apparatus is affected with a **central hearing loss**?

d   a. Outer ear   b. Organ of Corti   c. Middle ear   d. The brain

12.   High frequency sounds seem louder than low frequency sounds and consonants sound louder than vowels.

b   a. True   b. False

13.   Consonants carry most information in speech but are, for the most part, weaker sounds with relatively high frequency and are the most vulnerable to masking.

a   a. True   b. False

14.   Which of the following BEST describes the degradation effect of **masking** on the reception of spoken language?

        a.   Several frequency bands are cut out
        b.   Intermittent total cuts in speech
        c.   Increases the chance for expectation and other errors
        d.   Unwanted noise from the environment or radiomagnetic interference
e   e.   Both c and d apply to masking

Match the below listed characteristics of speech (a-d) with the appropriate statement in items 15 - 18
            (Use each only once)

        a. Intensity   b. Frequency   c. Harmonic composition   d. Speed

15.   Measured in Hertz (pitch) - Healthy human ear from 16 Hz - 20,000 Hz
b
16.   Length of pauses or time spent on the different sounds
d
17.   The quality, which can change meaning from sympathetic to sarcastic
c
18.   The volume - Measured in decibels   (The loudness)
a

19.   Which of the following BEST describes the degradation effect of **clipping** on the reception of spoken language?

        a.   The most effectively protection is to isolate or control it at the source
        b.   Increasing the volume of both the signal and the background noise will not enhance the message
        c.   Increases the chance for expectation and other errors
        d.   Several frequency bands are cut out
e   e.   Both c and d apply to clipping

20.   Hearing protection does not degrade speech intelligibility because the signal-noise ratio remains the same.

a   a. True   b. False

21.   Which of the following will **NOT** help reduce the phenomenon of expectation from leading to an error in understanding spoken language?

        a.   Wording positive and negative messages differently and place verbal stress on critical words
b   b.   Give both the expected and intermediate altitude in the clearance
        c.   Involve more than one crew member on important messages
        d.   Use data link with visual display
        e.   Readback confirmation or requesting a repeat if doubt exists

22.   What is the use of an articulation index good for?

a   a.   To compare the efficiency of different communication systems
        b.   To predict the voice level required for communication at certain noise levels
        c.   To measure the destructiveness of noise on the reception of speech
        d. To measure simultaneously the volume (intensity) and frequency (pitch) of speech

23.   Which of the following are problems associated with the development of automatic speech recognition (ASR) systems?

        a.   Difficulty in breaking human speech down into sequences of discrete words
        b.   Different people speak words differently (accents) and they so most are speaker-dependent
        c.   Requires many levels of complex processing
        d.   Harmonic composition or quality of speech can vary due to illness, fatigue, or emotion
e   e.   All of the above

24.   With regard to verbal confusion having a potential for causing accidents, in the 1986 survey of the
        NASA Aviation Safety Reporting System (ASRS) reports, what percentage involved some kind of oral
        communication problem?

c   a. 30%   b. 50%   c. 70%   d. 90%

Attitudes and Persuasion   6

01.   Attitudes are deep-seated characteristics which constitute the essence of a person.

b   a. True   b. False

02.   Which of the following would be the most resistant to change?

a   a. Personality   b. Attitude   c. Belief   d. Opinion

03.   Which of the below is defined by "A learned and rather enduring tendency to respond favorably or
        unfavorably to people, decisions, organizations or other objects"?

b   a. Personality   b. Attitude   c. Belief   d. Opinion

04.   Which of the below is defined by "A predisposition to respond in a certain way"?

b   a. Personality   b. Attitude   c. Belief   d. Opinion

05.   Which of the below does not necessarily infer a favorable or unfavorable evaluation?

c   a. Personality   b. Attitude   c. Belief   d. Opinion

06.   Which of the below is a verbal expression of an attitude or belief?

d   a. Personality   b. Attitude   c. Belief   d. Opinion

07.   Personality, attitudes and beliefs are intangible in as much as they cannot be seen or studied directly
        but only inferred from what a person says or does.

a   a. True   b. False

08.   Where do attitudes have their origins?

        a.   In early life experiences from the family
        b.   From Early life political orientation
        c.   In the social environment
        d.   The media
e   e.   All of the above

09.   Which of the below statements about the three components of attitudes is true?

        a.   Cognitive   - feelings held about it
b   b.   Affective   - feelings held about it
        c.   Behavioral   - knowledge, idea or belief about a subject
        d.   Affective   - what is said or done about it

10.   Which of the following is **NOT** a danger of stereotyping?

a   a.   It is justified when based on personal experience
        b.   It is frequently based on rumor or word of mouth
        c.   Often it is based on unsound generalizations or prejudices
        d.   It sets us up for the error of expectancy and we tend to see and hear what we expect

11.   Which of the below ways of measuring components of attitudes is the **best** for a researcher to use?

        a.   Measuring the strength of feeling
        b.   Measuring the degree of resistance to change
c   c.   Measuring the extent the attitude is acted upon
        d.   Measuring the degree or extent of occupation with it

12.   Which system of attitude measurement uses a scale of 1 (strongly approve) to 5 (strongly disapprove) with a list of opinions, each on a different aspect of a subject?

c   a. Thurstone Scale   b. Guttman Scale   c. Likert Scale   d. Herzberg Scale

13.   Which system of attitude measurement uses opinions placed in a sequence from one extreme to the other and you select the one you most agree with?

a   a. Thurstone Scale   b. Guttman Scale   c. Likert Scale   d. Herzberg Scale

14.   Which of the following statements concerning the effects of a group on the behavior of individuals in the group is **NOT** true?

a   a.   Usually the decisions made by a group will be more inhibited than those made by an individual acting alone.
       b.   An individual normally over-estimates their personal level of risk-taking.
       c.   Usually those who join a group feel the majority of the group share their own views
       d.   The group has a profound and extensive effect on the attitudes, beliefs, interests, behavior and even goals of the individual member.

15.   Which of the following statements concerning the effects of a group on the behavior of individuals in the group is **NOT** true?

       a.   Diffusion of responsibility allows the individuals in the group to share any adverse consequences.
       b.   Increased familiarity from group discussions increases familiarity with all aspects of the topic and provides confidence to accept more risk
       c.   There is experimental support for the assumption that risk is a socially desirable value and most social heros are risk-takers
d   d.   In a group compromise will eliminate the more risky alternatives
       e.   Something in the presence of others in the group weakens an individual's maintenance of socially acceptable norms of behavior

16.   Which of the following statements concerning the effects of a group on the behavior of individuals in the group is **NOT** true?

       a.   Group pressures to conform are less effective on an individual if an individual has confirmed the validity of their own attitudes and behavior.
b   b.   Usually the conformity of the group will reduce the dominance of a risk-taker leadership style and the result of the group action will involve less risk.
       c.   Group pressures to conform are less effective on an individual when they have a background of success in the subject or attitude in question.
       d.   Factors which constitute the influence of a group on the individual to maintain conformity often results in the distortion of perceptions, judgment and action and can have both positive and negative applications
       e.   When a decision is made by a group, it is likely to involve a greater element of risk than if it is made by an individual.

17.   Which of the following statements about the changing of a persons attitude is **NOT** true?

       a.   A person tends to dispute or reject any evidence presented that appears to contradict their strongly held attitudes.
       b.   Attitudes are more easily changed if they are not central to a belief or a part of one's basic perception of life and living.
c   c.   The persuasion technique of voluntary role reversal tends to help a person adopt a **more** intransigent position.
       d.   To increase a person's attitude strength elicit a personal commitment.
       e.   The inoculation method of exposing people to small doses of certain undesirable attitudes will increase resistance to change.

18.   A good training program is very likely to improve pilot personality traits.

b   a. True   b. False

19.   What is the most significant function of communication?

b   a. Instrumental   b. Persuasive   c. Informative   d. Ritual

408

20.   What is the most significant function of communication?

        a.   Obtaining something
        b.   Finding out or explaining
        c.   Part of a ceremony
d   d.   Modification of attitude or behavior

21.   What can be done to enhance the persuasive effectiveness of the message itself?

        a.   Show only one side of the argument if the audience does not already agree in principle or is
                more educated or intelligent.
        b.   Show both sides of the argument if the audience already agrees with you.
        c.   Use the law of recency if the first argument is more influential.
        d.   Use the law of primacy if the second argument is more persuasive.
e   e.   Include a conclusion if there are complex issues or the audience less bright .

22.   When considering the receiver in the persuasive communication process, It has been found that sometimes
        a degree of audience distraction during a message seems to **increase** the persuasiveness of the message.

a   a. True   b. False

Training and Training Devices   6

Match the below listed words (a - d) which BEST fits with the appropriate definitions in items 01 - 05.
        (Some may be used more than once)

        a. Education   b. Training   c. Instruction   d. Skills

01.   An organized and coordinated pattern of physical, social, linguistic or intellectual activity that
        leads to deciding on a course of action and carrying out that action.
d
02.   Activities associated with either education or training
c
03.   A broad-based set of knowledge, values, attitudes and skills suitable as a background upon which more
        specific job abilities can be acquired at a later stage
a
04.   A process aimed at developing specific skills, knowledge or attitudes
b
05.   It is seen as the precursor or foundation of training.
a

06.   Which of the following is best described by "An organized and coordinated pattern of physical, social
        linguistic or intellectual activity that leads to deciding on a course of action and carrying out that
        action"?

d   a. Education   b. Training   c. Instruction   d. Skills

07.   Which of the following statements about feedback is **NOT** true?

        a.   Feedback control means simply that the output of a system can regulate or control the input
b   b.   When there is no feedback present to regulate the input, the system is called a closed-loop
                system
        c.   The open loop system is full of potential difficulties.
        d.   In flying the pilot gets feedback mainly through instruments and outside cues.

08.   Which term is BEST used to label the condition where feedback available in the normal job situation and
        is native to the situation?

b   a. Negative transfer   b. Intrinsic   c. Positive transfer   d. Extrinsic

09.   Which of the following terms does your author use for labeling the process that involves informing the
        trainee when a signal is about to appear?

a.   a. Cueing   b. Guidance   c. Prompting   d. Pacing

10.   Which of the following terms does your author use for labeling the process that involves presenting the
        trainee with the correct response immediately after the stimulus?

c   a. Cueing   b. Guidance   c. Prompting   d. Pacing

11. Which of the following terms does your author use for labeling the process that involves physical control of control movements in the flare or formation training?

b   a. Cueing   b. Guidance   c. Prompting   d. Pacing

12. Which of the following has the greatest influence or control over the **learning** process?

a   a. Student   b. Training program   c. Instructor   d. Training situation

13. Which of the following has the greatest influence or control over the **training** process?

c   a. Student   b. Training program   c. Instructor   d. Training situation

14. In differentiating the various phases of learning which stage is BEST described by "involves perfection of performance, speeding it up and improving accuracy and precision"?

d   a. Cognitive   b. Practical   c. Associative   d. Autonomous

15. In differentiating the various phases of learning which stage is BEST described by "talking about the task and possible errors in a ground school"?

a   a. Cognitive   b. Practical   c. Associative   d. Autonomous

16. In differentiating the various phases of learning which stage is BEST described by "practice to reduce the errors while being accompanied by an instructor."?

c   a. Cognitive   b. Practical   c. Associative   d. Autonomous

18. Which kind of memory is described using the following characteristics?
- Time - Accurate recall within a few seconds
- Capacity - Six to eight items or chunks of information
- Processing - Limited level

c   a. cognitive   b. associative   c. short-term   d. long-term   e. autonomous

19. Which kind of memory is described using the following characteristics?
- Time - More time to accomplish
- Capacity - Storage space is no problem
- Processing - Preparation for storage is important

d   a. cognitive   b. associative   c. short-term   d. long-term   e. autonomous

20. Which of the following statements about memory loss is **NOT** true?

a
    a.   It is more effective to learn right after rather than right before sleep.
    b.   It is difficult to generalize about affected age because of the differences between individuals.
    c.   Usually it's not very dramatic until reaching the sixties.
    d.   Deterioration is selective - some memory holds up better than others.
    e.   Depression is a condition damaging to memory.

21. Which of the following statements about memory loss is **NOT** true?

    a.   It is difficult to generalize about affected age because of the differences between individuals.
    b.   Usually it's not very dramatic until reaching the sixties.
    c.   Deterioration is selective - some memory holds up better than others.
    d.   Depression is a condition damaging to memory.
e   e.   Mental function is not like physical which deteriorates with disuse.

22. In describing the difference between a training aid and training equipment the flight simulator would be a good example of a training aid because it provides for some active participation and practice.

b   a. True   b. False

23. Which of the following statements about training device fidelity is **NOT** true?

    a. Fidelity must be applied to each training situation taking into account the nature of the specific task being simulated.

    b. Research has shown that sometimes different degrees of fidelity have little impact on the trainee   or effectiveness of the training.

    c. Psychological fidelity depends upon the perception of the training devise by the individual student and is not necessarily dependent on either equipment or environment fidelity.

d   d. When only procedures are being learned, high fidelity in the equipment and the environment is preferable.

    e. Lack of certain kinds of fidelity can have negative training transfer

Match the below listed training systems (a - e) with the advantage or disadvantage that is BEST characterized in the statements that follow (24 - 28).   (use each only once)

    a. Tutorial
    b. Discussions
    c. Audio-visual methods
    d. Computer-based-instruction
    e. Lecture

24. It is non-participatory and modification and updating involves some skilled technical work.

c
25. No audience size limitation but encourages passivity and is hard to hold interest.

e
26. Is good for complex skills and for safety where training risks are involved.

a
27. Participation is strongly encouraged, has high interest level, but requires a skilled instructor to guide   to the objectives, and necessitates a limited size of participants.

b
28. Programs can be highly structured, have great flexibility, can incorporate testing and has good self pacing.

d
29. Which of the following statements about training device fidelity is **NOT** true?

    a. They should provide sufficient fidelity to obtain authorization from state certifying authorities for the simulators to be used as measures of the human proficiency.

    b. The degree of fidelity referred to in describing a simulator's performance means its accuracy or faithfulness with which a simulator reflects the real task

    c. It must be applied to each training situation taking into account the nature of the specific task being simulated.

d   d. Simulator features like freeze depart from true fidelity and thus reduce training effectiveness.

    e. The degree of fidelity may have little impact on training effectiveness.

30. When might **low fidelity** be appropriate for training?

    a. When a trainee must learn to make discriminations.
    b. When responses required are difficult to make.
    c. When responses are very critical to the operation.

d   d. When starting training or when only procedures are being learned.

06. The distinction between training aids and training equipment is that training aids provide for some form of active participation and practice by the trainee.

b   a. True   b. False

07. Research has shown that sometimes different degrees of fidelity have little impact on the trainee or effectiveness of the training.

a   a. True   b. False

08. Psychological fidelity depends upon the perception of the training devise by the individual student and is not necessarily dependent on either equipment or environment fidelity.

a   a. True   b. False

09. When only procedures are being learned, low fidelity in the equipment and the environment is preferable.

a   a. True   b. False

Documentation   6

01. Which of the following BEST describes the three basic aspects which require human factors optimism in all documentation?

      a.   Consistency, style of headings and references
      b.   Quality , appearance and readability
      c.   Technical accuracy, spelling accuracy and motivation power
d   d.   Written language, printing and layout

02. Who is better qualified to use the language and words of machines in technical documentation?

      a.   machinist
b   b.   human factors specialist
      c.   newspaper editor
      d.   speech therapist
      e.   literary scholar

03. Which of the following statements is true when writing for communicating with passengers?

      a.   The language used must be that of the reader
      b.   Understand the audience
      c.   Optimize for comprehension
d   d.   All of the above

04. Generally longer rather than shorter sentences aid comprehension.

b   a. True   b. False

05. Generally longer rather than shorter words aid comprehension.

b   a. True   b. False

06. Which of the following statements concerning the use of jargon in technical documentation is true?

      a.   It is the predominant specialized vocabulary of a technology.
      b.   Serves no useful purpose and should not be employed at all in technical documentation.
c   c.   It seems to be an unavoidable component of technical writing.
      d.   It motivates the reader who is new to the material and generates interest in the specialty.

07. Which of the following statements about the printing used in written documentation is true?

      a.   The form of letters and printing, and the layout have a significant impact on the comprehension of the written material.
      b.   The legibility of small type is greatly improved by increasing the space between the lines.
      c.   The sanserif type-faces have better legibility and are preferable for signs , placards or short material like checklists.
      d.   Even for headings it is better to use heavier type lower case letters rather than capitals.
e   e.   All of the above are true.

08. Which type of printed letters would be the minimum size for technical documentation?

b   a. 5 cpi   b. 10 point   c. 12 cpi   d. 15 point   e. 20 cpi

09. Which kind of type is best for legibility?

a   a. Sanserif   b. Serif   c. Italic   d. All capitals

10. Which kind of type is best for long passage comprehension?

b   a. Sanserif   b. Serif   c. Italic   d. All capitals

11. In typing use of underlining is better for headings than upper case.

a   a. True   b. False

412

12.   Italics are easier to read than normal type.

b   a. True   b. False

13.   Which of the following statements about the layout of technical documentation is **NOT** true?

       a.   The size and shape of the document is going to be the major factor for determining the layout.
       b.   Single column presentation is best for complex instructional material.
c   c.   Long text paragraphs should use spacing rather than indentation
       b.   Double column presentation is easier for speed and scanning.
       e.   Numbering of paragraphs is better than lettering.

14.   What are the functions of illustrations in technical documentation?

       a.   Motivate the reader
       b.   Help in recall from long-term memory
       c.   Aid in explanation
       d.   Help avoid jargon
e   e.   All of the above

15.   In constructing a questionnaire, which type of question is quick to answer and easier to process?

b   a. Open-ended   b. Closed-ended   c. Non-factual   d. Ambiguous

16.   In constructing a questionnaire, which type of question will usually provide wide information with
       little bias on the part of the respondent?

a   a. Open-ended   b. Closed-ended   c. Non-factual   d. Ambiguous

17.   Which of the following survey questions is the BEST example of violating the rule of biasing the
       participant by pleasing the questioner?

       a.   "Do you think that union bosses are properly reflecting the views of their members?"
b   b.   "Do you find our newly designed seats comfortable?"
       c.   "Do you enjoy traveling by road and rail?"
       d.   "When you are smoking do you always try to avoid causing discomfort to others?"

18.   In recent years the systematic and professional application of human factors to the development of
       aeronautical charts has been especially noteworthy.

b   a. True   b. False

19.   Which of the following statements about the development of aeronautical charts is **NOT** true?

a   a.   In recent years the systematic and professional application of human factors to the development
             of aeronautical charts has been especially noteworthy.
       b.   Generally the needs of the user have not had sufficient influence.
       c.   Research into human factors of charts is complicated and expensive.
       d.   What needs to be done to continue improvements in the human factors applications to charts and
             maps is a detailed task analysis and optimization for use on the flight deck.
       e.   Both a and b are false.

"Why Airplanes Crash"   5

01.   In the video "Why Airplanes Crash", what did John Nauce say lead to the Air Florida Flight 90 accident
       in 1982?

       a.   The crew's failure to have the ice removed from the wings
       b.   The Captain's failure to turn on the engine anti-ice
       c.   The First officer's failure to be more decisive with the lack of engine performance
d   d.   A chain of human failures

02.   What responsibility did the first officer have in the Air Florida crash of the Boeing 737 into the 14th
       Street bridge near Washington D.C.?

       a.   He pulled power on the wrong engine when there was an engine fire light.
b   b.   He knew something was wrong but was not forceful enough in expressing himself.
       c.   He failed to tell the Captain there was ice and snow on the right wing.
       d.   He talked the captain into taking off even though the captain had doubts about the performance
             of the engines.

413

03.   What does Dr Clay Foushee, the aviation psychologist in the video "Why Airplanes Crash", say might be parts of today's "right stuff" is in the cockpit?

       a.   A macho self sufficiency
       b.   A concern with the task (technical skills)
       c.   An ability to relate well to other people
       d.   All of the above (a,b & c)
e   e.   Only b & c above

04.   From the video "Why Airplanes Crash", What does Dr Clay Foushee see as one of the most useful techniques used as a part of the full mission simulation which is a part of the CRM training?

a   a.   The use of video replay
       b.   Overloading the crew with multiple emergencies
       c.   Using the feigned incapacitation of the captain to see if the first officer takes charge
       d.   The extensive role playing of the crew during the debriefing following the mission

05.   In CRM deficiency evaluation, shown in "Why Airplanes Crash", what, in addition to how well the crew handled the technical challenges, is also evaluated?

       a.   How well the emergency ended
b   b.   Their coordinated effort
       c.   How precisely they controlled the aircraft
       d.   How responsive the ATC system was

06.   From the video "Why Airplanes Crash", what recurring problems are illustrated by the disaster of Eastern Flight 401 which crashed in the Florida Everglades?

       1-   Captains that don't act as leaders and make decisions
       2-   Captains who don't set priorities, and delegate responsibilities
       3-   First officers or other crew who are unassertive
       4-   Complacent crew members

d   a. 1 & 2 only   b. 1,3,& 4 only   c. 1,2,& 3 only   d. All (1,2,3,& 4)

07.   According to the video "Why Airplanes Crash", what did the captain of Eastern Flight 401, which crashed in the Florida Everglades, neglect to do which allowed everyone else on the crew to be absorbed in the crisis?

       a.   Declare an emergency
       b.   Communicate the nature of the problem
c   c.   Divide up flying responsibilities
       d.   All of the above

08.   In the video "Why Airplanes Crash", what does Dr John Lauber say his point of view is about CRM training?

       a.   It is a refined method of training that is proven to be very effective.
b   b.   It's a concept still in development.
       c.   The training concept has received universal acceptance in the airlines.
       d.   Both a and c were stated.

09.   How does the Boeing 757 and 767 relive the pilots' mental workload according to the video "Why Airplanes Crash"?

       a.   They reduce the number of instruments the pilot has to monitor during an approach
       b.   They integrate data link instructions into the autoland mode of the automatic flight control
       c.   They use computer synthesized female voice to announce to the pilot that pre set performance parameters are being approached
d   d.   They use the computer to augment decision making

10.   What does Dr Wheeler say in the "Why Airplanes Crash" video is a very realistic concern the pilots have about the increasing use of automation in the cockpit?

       a.   They will loose their jobs
b   b.   The will have a loss of skills
       c.   The computer will make an unrecoverable error
       d.   Computers just can't fly as well as the human pilots

The number following the unit title is the number of questions needed from that unit to build a 50 question test # 1 performance objectives.

Displays   8

01.   Which of the following historical milestones in the development of cockpit displays was first?

        a.   Electronics and servo-driven instruments
        b.   Use of the cathode ray tube (CRT) for primary flight information
c   c.   A usable gyroscope for an artificial horizon
        d.   Serious attention given to the layout of the cockpit instruments

02.   Which of the following BEST describe the three necessary pillars upon which the progress of cockpit displays has been built?

        1-   Federal safety regulations
        2-   Avionics or aviation electronics
        3-   Ergonomics or human factors
        4-   Recommendations from accident investigations
        5-   Operational input

b   **a.** 1,3,& 4   **b.** 2,3,& 5   **c.** 1,3, & 5   **d.** 2,3, & 4   **e.** 3,4, & 5

03.   What did the Fitts, Jones and Grether studies contribute to the historical development of human factors in display design?

        a.   The development of servo-driven instruments
        b.   Pioneering the Cathode Ray Tube (CRT) in cockpit applications
c   c.   Used a more scientific approach to display design at Aero-Medical Laboratory in Dayton, Ohio
        d.   Encouraged airline development pilots communicating with manufacturers

04.   Which of the following areas of development in cockpit display design has had the most human factors studies?

        a.   Three pointer altimeter error studies
        b.   Fitts, Jones and Grether studies on display design
        c.   Scale reading accuracy studies
        d.   Studies of the effect of dial shape on legibility
e   e.   Studies on the human factors of CRTs or visual display terminals (VDTs)

05.   Which of the following BEST describes how aircraft **displays** fit into the SHEL model for human factors interface?

a   a.   Hardware   --->   Liveware
        b.   Software   --->   Liveware
        c.   Liveware   --->   Software
        d.   Liveware   --->   Hardware

06.   Which of the following BEST describes how aircraft **controls** fit into the SHEL model for human factors interface?

        a.   Hardware   --->   Liveware
        b.   Software   --->   Liveware
        c.   Liveware   --->   Software
d   d.   Liveware   --->   Hardware

07.   Which of the following is true about the first major human factors problem in the mission of display technology?

        a.   Human sensory capacity is very limited.
        b.   Human information transmission rate is enormous.
        c.   Information from sensory input does not need to be filtered, stored and processed.
d   d.   Display design must present the information in a way to help the brain process it.

08. Which of the following is **NOT** true about the first major human factors problem in the mission of display technology?

    a.   Human sensory capacity is enormous.
    b.   Human information transmission rate is very limited.
c   c.   Information from sensory input does not need to be filtered, stored and processed.
    d.   Display design must present the information in a way to help the brain process it.

09. Which instrument is the BEST example of a **qualitative** display classification?

    a.   Attitude director indicator (ADI)
b   b.   Turn coordinator or slip and turn indicator
    c.   Fire warning light
    d.   Landing gear indicator
    e.   Horizontal situation indicator (HSI)

10. Which instrument is the BEST example of a **systems status** display classification?

    a.   Attitude director indicator (ADI)
    b.   Turn coordinator or slip and turn indicator
    c.   Fire warning light
d   d.   Landing gear indicator
    e.   Horizontal situation indicator (HSI)

11. When comparing aural with visual display design visual are usually omnidirectional while aural are usually not omnidirectional.

b   a. True   b. False

12. Human factors study has shown that analog is better for direction or trend while digital is better for accuracy?

a   a. True   b. False

13. Which way should the mechanical drum rotate to display the next higher numbers according to the findings of human factors research?

b   a. Up   b. Down

14. What is happening to the mechanical and electro-mechanical alphanumeric displays recently?

    a.   They are becoming the predominant type in the modern cockpit.
    b.   Extensive human factors research and modifications are taking place.
c   c.   They are being replaced by electronic displays.
    d.   Both a and b are correct

15. Which of the styles of displaying data on scales has the **least** chance for errors?

e   a. Semi-circular   b. Vertical   b. Horizontal   d. Circular   e. Open window

16. Which of the styles of displaying data on scales has the **greatest** chance for errors?

a   a. Semi-circular   b. Vertical   b. Horizontal   d. Circular   e. Open window

17. Which of the following human factors principles to consider in designing display scales is **NOT** correct?

    a.   Avoid varying progression if possible.
    b.   Single unit progression is best.
    c.   Eliminate the decimal points.
d   d.   The graduation base is usually on the inside with major markers extended outward.
    e.   The pointer tip should just touch the tip of small graduations.

18. Which of the following human factors principles to consider in designing display scales **is** correct?

    a.   Multi unit progression is best.
    b.   Include the decimal points in the scale to avoid confusion.
    c.   The graduation base is usually on the inside with major markers extended outward.
d   d.   Sunburst design has the graduation base on the outside.
    e.   The pointer tip should cover the small graduations.

19.   Which of the following statements about the application of cathode ray tube (CRT) technology in cockpit displays is **NOT** true?

    a.   The Electronic Attitude Director Indicator (EADI), developed by Boeing, is used as a flight instrument display.

b   b.   The Primary Flight Display (PFD), developed by Airbus, uses two CRTs for a warning and a systems display.

    c.   The Electronic Centralized Aircraft Monitoring (ECAM), developed by Airbus provides a warning display and a systems display.

    d.   The Engine Indicating and Crew Alerting System (EICAS), developed by Boeing was designed to provide systems and warning information.

20.   Which of the following statements about the application of cathode ray tube (CRT) technology in cockpit displays **is** true?

    a.   The Electronic Attitude Director Indicator (EADI), developed by Boeing, is used as a flight instrument display.

    b.   Airbus uses a total of three CRTs; one for the Primary Flight Display (PFD) and two for a warning and systems display.

    c.   The Electronic Centralized Aircraft Monitoring (ECAM), developed by Airbus provides a warning display and a systems display.

    d.   The Engine Indicating and Crew Alerting System (EICAS), developed by Boeing was designed to provide systems and warning information.

e   e.   All of the above are true.

21.   A **fail-passive** automatic landing system is one designed so when a failure occurs there is sufficient redundancy that an operational capability remains to touchdown and the pilot does not need to take over.

b   a. True   b. False

22.   A **fail-operational** automatic landing system is where the autopilot simply hands over the aircraft in a steady condition without significant aircraft deviation or an out of trim condition.

b   a. True   b. False

23.   What are the three **fundamental objectives** for application in the design of all flight deck warning systems?

    1- Alert   2- Cue   3- Report   4- Guide   5- Advise   6- Instruct

a   **a.** 1, 3, & 4   **b.** 1, 3, & 5   **c.** 2, 4, & 6   **d.** 2, 3, & 4   **e.** 3, 4, & 6

24.   What are the four **functional classes** of the design of cockpit warning systems?

    a.   Alert, report, guide and cue

    b.   Emergency condition, abnormal condition, advise and alerts

    c.   Sight, sound, feel and smell

d   d.   Performance, configuration, systems and communications

Controls   3

01.   Which of the following BEST describes how aircraft controls fit into the SHEL model for human factors interface?

    a.   Hardware ---> Liveware

    b.   Software ---> Liveware

    c.   Liveware ---> Software

d   d.   Liveware ---> Hardware

02.   Which of the controls listed below normally requires the lowest control force on the part of the pilot to actuate?

a   a. Electrical   b. Hydraulic   c. Mechanical   d. Pneumatic

417

03.   Which primary design principle for controls is being used when considering "feel" and smoothness of control movement?

        a.   Coding
        b.   Direction of movement
c   c.   Resistance
        d.   Control-display distance ratio

04.   Which primary design principle for controls is being used when considering improved identification and reduction of errors and time for selection?

a   a.   Coding
        b.   Direction of movement
        c.   Resistance
        d.   Control-display distance ratio
        e.   Inadvertent actuation protection

05.   Which primary design principle for controls is being used when the size shape or color of different controls is unique?

a   a.   Coding
        b.   Direction of movement
        c.   Resistance
        d.   Control-display distance ratio
        e.   Inadvertent actuation protection

06.   Which primary design principle for controls is being used when the control uses gating or is lever-locked?

        a.   Coding
        b.   Direction of movement
        c.   Resistance
        d.   Control-display distance ratio
e   e.   Inadvertent actuation protection

07.   Which of the following is **NOT** a correct human factors principle used for the design of a cockpit keyboard control?

a   a.   Speed is better than accuracy
        b.   Must be able to use in turbulence or darkness
        c.   May be in a less than optimal location or space
        d.   Usually will utilize only one hand

Space and Design   3

01.   What does accounting for movement of body parts and the forces which they can apply refer to in the space and design of the cockpit or cabin?

d   a. Anthropometry   b. Anthropometer   c. Psycosociology   d. Biomechanics

02.   What does the study of human dimensions of size, weight, stature, seated eye height and reach have to do with?

a   a. Anthropometry   b. Anthropometer   c. Psycosociology   d. Biomechanics

03.   In considering human size dimensions it is important to know there are differences in nationalities and ethnic groups as well as the fact that average size is increasing.

a   a. True   b. False

04.   Other differences which must be considered by designers of cockpit equipment for worldwide use is ethnic differences such as the average asians are shorter overall with a shorter trunk and longer legs compared to Europeans.

b   a. True   b. False

418

05.   Which of the following is true concerning general Human Factors guidelines?

        a.   Small people determine clearance.
        b.   Large people determine limits for reach.
c   c.   The design must usually provide for a range of adjustment .
        d.   Percentile is a means of expressing the average of a range of sizes to be accommodated.

06.   Which of the following statements about designing better cockpits for the human operator is true?

        a.   Designers were lacking in an overall view of the flight crew
        b.   Designers were lacking formal human factors training
        c.   It's much better to design out the human factors problems in the conceptual stage rather than to find it in the development stage and then have to fix it
d   d.   All of the above are true
        e.   Only b and c above are true

Cockpit Layout   4

01.   Which of the following are factors which influence the design of the aircraft windshield?

        a.   Downward visibility requirement
        b.   The location of the "design eye" position
        c.   The requirement for pedestal panel space
        d.   All of the above
e   e.   Only a and b above

02.   Which of the following statements about designing for the space between the pilots is **NOT** true?

a   a.   Wider spacing gives more inboard access
        b.   Closer spacing gives better cross-monitoring
        c.   Closer spacing has a greater loss of pedestal panel space
        b.   Wider spacing gives better outside lateral visibility
        e.   Wider spacing gives better More use of pedestal panel space

03.   What is the typical viewing distance from the pilot's eye to the main instrument panel?

b   a. 65-70 cm (26-27 in)   b. 71-78 cm (28-31 in)   c. 79-93 cm (31-37 in)

04.   How close might some overhead panels be to the average pilot's head?

d   a. 65 cm (26 in)   b. 40 cm (15 in)   c. 35 cm (13.5 in)   d. 20 cm (7.9 in)

05.   Flight guidance control panels are now generally mounted on the main instrument panel.

b   a. True   b. False

06.   Design considerations for system quantitative information indicate that pointers with extended tails are better than short ones.

a   a. True   b. False

07.   Which principle for toggle switch design would be applied if an overhead control panel had a switch that was off in the aft facing position?

a   a. Forward-on   b. Sweep-on   c. Overhead-aft-on   d. Inboard-on

08.   Which of the following recommendations for cockpit seat design is **NOT** correct?

        a.   Nagging back pain or discomfort may effect pilot behavior, motivation and performance.
        b.   An important result of a poor sitting posture is an incorrect curvature of the lumbar region of the spine.
c   c.   Seat height is critical for comfort and should be adjustable through a range of 4 inches.
        d.   The seat cushion should allow adequate ventilation, be floatable and flame resistant.

01.   What are the two major problems faced in trying to optimize application of human factors in the cabin between the passengers and the crew?

        1-   safe smoking area
        2-   comfortable working area
        3-   smoke free area for non-smokers
        4-   an atmosphere of reduced anxiety
        5-   passenger rest area
        6-   adequate emergency procedure familiarization

b   a. 1 & 3   b. 2 & 5   c. 1 & 3   d. 2 & 3   e. 1 & 4

02.   What should be the primary motivation for the flight attendant?

        a.   The opportunity to travel
        b.   The experience of working with aviation oriented people
c   c.   The satisfaction of working with people
        d.   The high pay and benefits of an airline job

03.   What is the **major** cabin region demanding Human Factors attention in matching the hardware to the liveware to maximize safety and efficiency?

        a.   Lavatory
b   b.   Galley
        c.   Passenger seats
        d.   Overhead luggage racks
        e.   Emergency exits

04.   What does delethalization refer to in the design of cabin facilities or equipment?

        a.   Process of removing sharp objects or protrusions
        b.   Insuring adequate retention of loose equipment
        c.   Prevention of injury from high-jacking and terrorism
d   d.   Only a and b above
        e.   All of the above (a,b and c)vice dependent upon?

        a.   General space and layout
        b.   Number and location
        c.   Control panel design
d   d.   All of the above

05.   What does the term "delethalization" refer to in designing for human factors in the cockpit and cabin?

        a.   Process of removing sharp objects or protrusions
        b.   Insuring adequate retention of loose equipment
        c.   Use of anti-misting fuels to reduce fire and lethal smoke after crashing
d   d.   Only a and b above
        e.   All (a, b & c) of the above

06.   What special human factors considerations must be made when designing emergency equipment?

        a.   May need to be used in total darkness
        b.   May need to be operated by passengers unfamiliar with it
        c.   Environment may be chaotic
        d.   Human behavior under conditions of stress
e   e.   All of the above

07.   Which best describes the major hazard in survivable aircraft crashes?

a   a.   The toxic smoke and fumes of fire
        b.   Thermal injury from post crash fire
        c.   Crash deceleration forces in excess of human tolerance
        d.   Malfunctioning emergency equipment
        e.   Drowning and hypothermia after escaping from the aircraft

08. What is seat pitch in the design of cabin passenger seats?

      a. The angle which the seat is reclined
b   b. The spacing between the seats (leg room)
      c. The maximum pitching G forces the seat is designed to withstand
      d. The spacing between the seats (width of the seat pan)

09. What increase in load force is derived from upper body restraint in an aircraft seat?

      a. The load force can be half again as high
b   b. Forward load tolerance can be doubled
      c. The forward load force can be three times higher
      d. Loads as high as 83 Gs can be sustained

10. Which of the following is the BEST example of a **Service Duty** broad category which cabin staff must perform for passengers?

      a. Duties determined by regulation
      b. Administrative paperwork for the airline
c   c. Duties established by the marketing department
      d. Checking and preparation of flight safety equipment

11. Which of the following is the BEST example of a **Operational Tasks** broad category which cabin staff must perform for passengers?

a   a. Duties determined by regulation
      b. Duties established by the marketing department
      c. Meal provisions
      d. Interaction with passengers

12. Who in the cabin seems to suffer more injuries from in-flight turbulence?

      a. Sleeping flight attendants
      b. Passengers in the lavatory
      c. Intoxicated passengers
d   d. Working flight attendants

Make the **BEST** match for appropriate factors of concern in improving the interface between the L-E components of the cabin listed below (a -e) with example given in each of the following items 13 - 17.
(Use each factor listed only once.)

      a. Noise
      b. Temperature
      c. Ozone
      d. Smoking
      e. Circadian and time zone effects

13. It is one of the main causes of emotional friction in the cabin and also requires additional maintenance.
d
14. It appears that the damage from this factor is far more related to the concentration level than the period of exposure.
c
15. Passengers notice this increases with speed an is less along the aircraft center-line than nearer the sides of the fuselage.
a
16. The most significant environmental area which involves a conflict between the needs of the cabin attendants and the passengers.
b
17. Scheduling of various cabin activities such as serving of meals or movies should consider this factor.
e

18. Which of the following statements about cabin pressurization is **NOT** true?

      a. Normal cabin pressure is maintained between 5,000 - 8,000 ft.
      b. Time of useful consciousness (TUC) depends on aircraft altitude, rate of decompression an activity of the occupant.
c   c. Time of safe consciousness (TSC) is the time an occupant, denied adequate oxygen, will be able to put on the oxygen mask safely.
      d. Flight and cabin crew must be on supplemental oxygen before TUC expires.

01.   Crew-to-passenger communications are enhanced by the easy to design and maintain public address (PA) system in most aircraft.

b   a. True   b. False

02.   FAA regulations prohibit an airline from allowing on board any person who appears to be intoxicated and require a breath test for questionable cases.

b   a. True   b. False   .

03.   Which phase of flight seems to be the most difficult for passengers with fear of flying or flying phobia?

a   a. Takeoff   b. Climb   c. Maneuvers   d. Landing

04.   What other fears might have to be treated along with the fear of flying?

        a.   Fear of heights (acrophobia)
        b.   Fear of fire (pyrophobia)
        c.   Fear of closed (tight) places (claustrophobia)
        d.   Fear of speed (accelaphobia)
e   e.   Usually only a and c above

05.   Which of the following might be used as a practical measure that can be taken against passenger abuse/assault of flight crew?

        a.   Warn a passenger of in-flight off-loading (without a parachute)
        b.   Be trained on the location and use of restraining equipment
        c.   Take the passengers name and obtain a list of witnesses
d   d.   Only b and c above
        e.   All of the above (a, b and c)

06.   What important Liveware-Liveware principles are highlighted from feelings flying passengers have about their situation in the aircraft cabin?

        a.   Passengers look to the crew for leadership, guidance and instructions to insure escape, survival and rescue.
        b.   A group becomes more dependent on a leader in stress situations.
        c.   The PA systems or megaphones may be the only way to provide guidance to passengers in an emergency.
        d.   Only b and c above
e   e.   All of the above

07.   Which of the following is true of the accident survivability statistics which come from analysis of human behavior in aircraft emergencies?

a   a.   Seating close to an emergency exit will enhance chances
        b.   The very young and very old are less vulnerable
        c.   Females are more likely to survive
        d.   Only a and c are true
        e.   All (a, b & c) of the above are true

08.   Life-vest design has remained virtually unchanged in the last 30 years even though FAA tests in 1984 showed one in three passengers failed to don their life-vests successfully, even after viewing a demonstration.

a   a. True   b. False

Education and Applications   6

01.   What seems to be the main difficulty in applying human factors principles in the aerospace industry?

        a.   There is not enough researched knowledge in the field.
        b.   The designer, manager or operator is not concerned with practical applications of human factors.
c   c.   The academic knowledge in the field is not being translated into operationally usable language.
        d.   There is not enough economical pressure to motivate managers to implement human factors principles.

02.   What background seems to be the most common for human factors specialists?

        a.   Medical training in aviation physiology
b   b.   Engineering, industrial or experimental psychology
        c.   Systems design safety
        d.   Clinical psychiatry
        e.   NTSB or ICAO human factors accident investigation

Match the below listed positions (a - d) with the educational requirement expressed in items 02 -05 that **BEST** fits the needs of the position.   (Use each position only once.)

        a. All staff   b. Supervisors   c. In-house specialist   d. Consultant

03.   Requires a university higher degree and industry research
d
04.   Requires a compressed 2-week course
b
05.   Requires a university degree in an appropriate field
c
06.   Would be the best place to apply the KLM human factors awareness audio-visual units
a

07.   Which of the following is **NOT** what the CRM or LOFT programs are designed to improve through specific human factors education or training applications?

        a.   leadership and communication
        b.   task distribution and priorities setting skills
        c.   ability to monitor information sources and individual performance
d   d.   change personality traits

08.   Which of the following is **NOT** one of the elements of Cockpit Resource Management (CRM) training?

        a.   Leadership
        b.   Communication
c   c.   Personality adjustment
        d.   Task distribution
        e.   Setting priorities

Select from the list of agencies listed below (a - d) the one which BEST fits with the descriptions of responsibilities in items 09 - 12. (use each only once)

        a.   Designer
        b.   Manufacturers
        c.   The customer or operator (Airline)
        d.   The state certifying authority

09.   Who is responsible for the initial training on a new aircraft?
b
10.   Who has responsibility for designing human factors into the basic operating procedures and documentation?
a
11.   Who approves the training program of the operator?
d
12.   Who has the major responsibility for the conduct and cost of training on an operational aircraft?
c

13.   What is the key to success in human factors?

        a.   Government funding
b   b.   Education
        c.   Research and testing
        d.   Thorough mishap investigation

01.   Which of the following is a **Human Performance** issue in human factors investigation?

a   a.   Physiological concerns
     b.   Cockpit design concerns
     c.   Operations concerns
     d.   Life Support and Personal Equipment concerns

02.   Which of the following is a **Environmentally Oriented** issue in human factors investigation?

     a.   Physiologic and Biodynamic concerns
     b.   Biomechanical concerns
     c.   Psychological concerns
d   d.   Egress and Survival concerns

In items 03 - 07 match the below listed human performance or environmentally oriented human factor investigation concern with the appropriate example.

     a.   Psychological
     b.   Life Support and Personal Equipment
     c.   Physiologic and Biodynamic
     d.   Cockpit Design
     e.   Egress and Survival

03.   Assessing the aircraft crashworthiness for comparison with human impact tolerances.
e
04.   Evaluating how perception, training, or psychomotor capabilities were involved.
a
05.   Investigating how seat position, visibility, or switch and control locations might have affected the mishap.
d
06.   Investigating to see if hearing or visual impairment may have contributed to the mishap.
c
07.   Evaluating how cockpit environmental control or oxygen delivery system may have affected the mishap.
b

In items 08 - 12 match the below listed human performance or environmentally oriented human factor investigation concern with the appropriate example.

     a.   Psychosocial
     b.   Facilities and Services
     c.   Biomechanical
     d.   Operations concerns
     e.   Institutional, Personnel, and Management

08.   Was the pilot distracted or under stress by policies and issues of promotion.
e
09.   During the spin recovery the pilot could not apply full rudder because the legs were too short.
c
10.   The pilot was trying to live up to the "macho" image of the pilots he wanted to associate with.
a
11.   What were the special flight stresses that may have been present due to the nature of the mission.
d
12.   The pilot was not provided access to adequate nutrition at either of the stopovers during nine
b   hours.

13.   Which of the following is NOT an example of the psychological human factors general problem areas?

     a.   Training
b   b.   Supervisory influences
     c.   Perception and attention
     d.   Perceived stress and coping styles

14. Which of the following is NOT an example of the institutional, training or management issues that can have an impact on human factors?

    a. Internalization of unit and organizational values
    b. Policies and issues of evaluation and promotion
c   c. Mission demands and special flight stresses
    d. Selection policies

15. Which of the following is NOT an example of the <u>life support and personal equipment</u> aspects of human factors the safety investigator needs to examine for accident cause?

a   a. Assessing aircraft crashworthiness for comparison with human impact tolerances.
    b. Checking cockpit and cabin environmental control to eliminate or confirm cause.
    c. Examining the oxygen delivery system to verify it functioned properly during a pressurization loss.
    d. Investigating the protection provided to the aircrew by them wearing fire resistant uniforms.

<u>Video "The Wrong Stuff"</u>   <u>4</u>

01. What does the video "The Wrong Stuff" focus on?

a   a. Flight crew behavior
    b. Failures of the ATC system
    c. Problems of new technology
    d. Modern weather problems

02. In the famous incident of the 747 at Nairobi cited by Roger Green, why did the crew choose not to believe the warnings about being below the glide slope?

    a. Visual cues verified the warnings were false
    b. There was disagreement among the crew
    c. The crew was distracted by an engine failure
d   d. They did not fit into their model

03. What did John Lauber point out the crew failed to do when reviewed the crash of the 707 into the mountain on Barla Indonesia while attempting an NDB approach?

    a. recognize station passage
    b. set in the correct altimeter setting
c   c. positively resolve the ambiguities
    d. verify an obviously wrong assigned altitude by ATC

04. During the Boeing 727 simulation with a No 3 engine fire, what error in the cockpit resource management did the captain make relative to the flight engineer?

    a. gave him the wrong configuration
b   b. overloaded him with tasks
    c. did not accept his vital inputs
    d. sent him back in the cabin to fight the fire

05. According to Bob Helmreich, which of the 3 components at the test pilot right stuff creates problem when trying to function as an effective team?

    a. High technical competence
    b. Rugged individualism
    c. High level of competitiveness
d   d. Both b and c above

06. When captains in a simulator study pretended to be incapacitated, what percent of the simulator crashed when the copilots failed to take over?

d   a. 10%   b. 15%   c. 20%   d. 25%

07. What did Peoples Express focus on the eliminate the wrong stuff from their cockpits?

    a. extensive use of cockpit resource management resource
b   b. selection screening for people who can work with people
    c. line of flight training (LOFT) for flight and cabin crew
    d. selection of good looking people with good technical skills

425

08.   What has drastically changed the nature of the pilot's job in modern transport aircraft?

a   a.   computer automation
        b.   higher performance engines
        c.   wide bodied aircraft
        d.   using only 2 pilot crews

09.   What does the "Electronic Cocoon" do to help strike the balance between man and machine?

        a.   allows the crew more freedom to operate
        b.   provide warning to the crew when they approach limits
        c.   create a dialogue between man and machine
d   d.   all of the above

Video 'Top Gun and Beyond   4

01.   What are the engineers of today's fighter aircraft not adequately dealing with?

c   a. Psychology   b. Technology   c. Biology   d. Statistics

02.   At the speeds of today's modern fighter aircraft how much time is available for a fighter pilot to deal
        with the enemy who is 20 miles away?

b   a. 15 sec   b. 30 sec   c. 1 min   d. 2 min

03.   What was the result of all the various inputs the pilots of the Vietnam era had to deal with?

a   a.   Pilot saturation and overload
        b.   Cockpit display design improvements
        c.   Peak pilot psychological performance
        d.   The best designed aircraft ever developed

04.   What happens to a fighter pilot if he does not unload and get the blood back in his head after
        experiencing tunnel vision and first blackout?

        a.   He sees black and white dots.
        b.   He experiences "red out".
c   c.   He experiences a loss of consciousness.
        d.   He has heart failure.

05.   What did the video "Top Gun and Beyond" give as the estimate of pilot deaths in 5 years from "G" loss
        of consciousness?

d   a. 5   b. 10   c. 15   d. 20

06.   According to the video "Top Gun and Beyond", how long does it take for the higher learning, cognitive
        centers of the brain to start working adequately after a "G" loss of consciousness incident?

c   a. 12 sec   b. 30 sec   c. 2 min   d. 10 min

07.   How many extra "Gs" of tolerance can the anti "G" straining maneuver bring the pilot?

b   a. 1   b. 3   c. 5   d. 8

08.   What new study is being conducted to help the aircraft monitor the status of the pilot?

a   a.   Brain wave monitoring
        b.   Blood pressure monitoring
        c.   Eye movement monitoring
        d.   Breathing monitoring

09.   What are the "smart systems" being developed at Wright-Patterson AFB designed to help the pilot with?

        a.   elevated brain blood pressure
        b.   better instrument scan patterns
c   c.   greater situational awareness
        d.   improved aircraft control inputs

426

10.   With all the technological developments, why can't the computer replace the pilot?

a   a.   the human flexibility advantage
       b.   the computer visual deficiency
       c.   the computer lack of processing speed
       d.   the lack of computer accuracy

a   a.   the human flexibility advantage
       b.   the computer visual deficiency
       c.   the computer lack of processing speed
       d.   the lack of computer accuracy

Printed and bound by CPI Group (UK) Ltd, Croydon, CR0 4YY

23/10/2024

01778254-0020